Darwin
or
Adam and Eve

Revealing the Evidence!

By

Stan Bachelder

Allow me to Think ~ is it Evolution or Creation?

Printed in Canada

ISBN 978-0-9940138-0-4

FIN 17 07 2015

Library and Archives Canada Cataloguing in Publication

Bachelder, Stan, 1936-, author
    Darwin or Adam & Eve : revealing the evidence / Stan Bachelder.

ISBN 978-0-9940138-0-4 (pbk.)

    1. Creationism. 2. Evolution. 3. Religion and science. 4. Bible and evolution. I. Title. II. Title: Darwin or Adam and Eve.

BS651.B23 2015   231.7'652   C2015-901903-6

Table of Contents

About the Author...

Stan Bachelder is the son of Arthur and Myrtle Standish-Bachelder, born in Rougemont, Québec, Canada, the youngest in a line of three brothers and two sisters, now all deceased. Three sons and three daughters call him father with many grandchildren and great-grandchildren.

Rougemont was the location of his first formal education in a one-room schoolhouse, then later in Montréal and Granby.

NCR, National Cash Register Company provided an excellent business education in their accounting machine and computer systems division. He successfully held a Senior Representative position with this company in Montréal and Toronto for several years. He invented one of the first electronic cash registers in a company he established with a partner.

He started businesses called Simplified Operating Systems, and Info Management Corporation, operating in Canada and the United States, consulting and installing integrated computer systems for a wide variety of companies. The powerful personal computer replaced these older mini computers making integrated processing much easier.

Stan did the initial research obtaining the detailed information needs of these companies before programming and installing these computer systems.

He is a descendent of United Empire Loyalists and related to a pioneer called Stephen Bachiler of Hampton, New Hampshire. This connection to a very early pioneer gives him many relatives in Canada and the United States.

Stephen sailed to the New World in 1632 from the city of Southampton, a non-conformist Puritan vicar from Wherwell on the River Test, in Hampshire, England.

He started a colony in Winnacunnet, a name he changed to Hampton, in New Hampshire with many followers, founding a Congregational Church there in 1638 at the age of seventy seven years. This church boasts one of the oldest continuous congregations in the United States. A park is located in that town dedicated to his memory and to the families that settled there with him at a time when that area was only wilderness.

The non-conformist thinking principles expressed in this book are similar to that of the author's ancestor, Stephen Bachiler. Stan Bachelder includes many questions encouraging his readers to "think" rather than simply accepting any unproven theories. The front cover shows Auguste Rodin's "The Thinker" as part of this theme. He uses personal experiences and a skill in doing research, rather than someone's opinion, comparing evolution with creation, in an attempt to dispel false premises. He challenges his readers to use the conclusions of his research, and the many references supplied, in their search for the truth.

Stan is now retired, residing with his wife Gillian on a farm in the Caledon area, just northwest of Toronto in Ontario, Canada.

Preface and Acknowledgments

What is required to write a book? I explain some of the factors that influenced me.

I mention a lawyer in one chapter, rambling on with that lawyer about a problem I had while he listened attentively. Then, in one sentence, he would state my case! I am most grateful to my wife Gillian for her loving patience, listening to rambling descriptions of some topic involving this book. She would then summarize it all with one sentence. This can be most disturbing, but what a great blessing she is for me!

Family and friends listened patiently over nearly ten years or more while I told them I was writing a book. One friend was most helpful, as he looked right at me saying, "We often hear about that book but when will it be finished?" He inspired me to "get on with it." I thank my family and friends who encouraged me to write this book. A special thanks to Frances Schatz, my Editor, for her dedication and patience in working with me.

In my career of installing business computer systems, I am grateful for many talented programmers. I would find them trying to solve program problems at night after business hours. I remember telling them to go home, sleep on it; you will have the answer in the morning. It worked!

In the course of writing this book, I had to learn to stop the chattering in my head, learning to listen to that small voice within. I often rose early in the morning having the answer to a question so troubling the night before or even during the night. Do we need to shut out all the noise around at times, "stopping" to listen to that wee small voice within?

While on vacation in Mexico and Florida, I met strangers who also encouraged me. These strangers who are now friends encouraged me to include the following story that I related to them.

This is the story. There was a time when I was discouraged and not very pleased with anything that I had written. I considered finding a University where Professors could teach me how to write a book. A magazine appeared in my mailbox, not addressed to me, rather to a close neighbour. The mail carrier made a mistake. I looked through that magazine which described home courses by talented university professors on how to write a book, along with several other helpful courses. I took note of this before returning the magazine to its rightful owner who proceeded to tell me she did not want the magazine. She encouraged me to keep it. I admit not remembering all the advice provided in the course but my "word processor" stopped objecting so much to my writing. Was this just a coincidence?

The front cover of the book has a picture of "The Thinker" for a special reason only suggesting that in an age of much influence that you and I stop to take time for our own independent thought.

Introduction to a Balanced Outlook

Personal experience, a valid scientific principle of discovery, and research over a considerable period, lead me to conclude that Mind and Body exist on one side; Spirit and Soul exist on the other side of a balanced outlook on life. We live in a world that is natural, finite, and complete with a very necessary set of laws and rules. These laws govern what we do, rules being very necessary for us to learn. These laws cannot be in constant change or our very existence on this planet would be unmanageable.

However, only considering this Natural World can lead us to knowledge with a dead end, no conclusion. In addition, by considering only the Natural World, we limit our access to knowledge, coming to conclusions that may not be logical. Humans progress from knowledge; they do not inherit this knowledge over time, as evidenced by our need to teach each new generation what we have learned.

The Spiritual world is external to our world; to us it is eternal, requiring a completely different set of laws, and rules, of which our natural existence has none, or little idea. If one is brave enough to consider, we discover a glimpse of another dimension, a spiritual dimension, another world. If we choose to consider this other dimension, we look dimly at an existence of Spirit, a form of existence that is indeed foreign to us. It gives us answers that we will never find if we resolve to consider only this finite Natural World. There is evidence of this spiritual dimension presenting itself to us on certain occasions, recorded for us to explore. These supernatural occurrences are undeniable, historically accepted as truth. The application of our thought to them provides a logical description, a beginning for our existence, complete with a logical end, then reason for our existence. If we choose to ignore them limiting our knowledge to a Natural Worldview only, we are missing out on the benefits of a much larger story.

The following series of chapters, based on my life of practical experience, is yours to consider. As I did my research, I found many others coming to similar conclusions.

Who may profit from this book? Who may benefit from this knowledge?

The Student, learning only about natural evolution without question, hearing only one point of view in grade school, then later in university.

The Spiritual Believer who may question one's personal belief but needs more information.

The Non-Spiritual Person can be interested, someone who considers only natural events.

The Religious Person may be someone who wants to believe the Genesis story, but needs a way to prove it.

The Liberal Minded Person, believing the "Gap, Deep Time," or other theories, thinking you can combine the idea of evolution with the principle of creation, two opposites.

The Compromising Person who may believe there is no conflict between creation and evolution.

The Accepting Person may be convinced that evolution is a fact only because many scientists claim it to be so.

The Uninformed Person is someone who may need to know the source of the Evolution Theories and the reasoning behind them.

This story is dedicated to the person with a brave questioning mind; a person who searches for the truth; a person who loves the truth; a person who chooses to apply and benefit from the truth.

The choices we make determine our destiny; choose well my friends!

Chapter One

Red Flags

Curiosity, was this emotion getting aroused again, leading me on yet another quest for answers?

It was July 20, 1969 when on that day man reached the moon traveling on space ship Apollo 11. Headlines around the world proclaimed man's first walk on the moon, Neil Armstrong stating for all to hear, "That's one small step for man, one giant leap for mankind."[1]

We observed the effect of that small step, an astronaut's boot sinking less than an inch in moon dust leaving a distinctive impression on the moon's surface. Remembering some comments over the radio, I recall someone to say, "We now know our moon cannot be more than 12,000 years old, not enough dust." Never forgetting that comment, I decided to get more information from a knowledgeable person on such a subject.

Many years passed before finally meeting Sam* an interesting Geologist, now 80 years old, retired from most of his life at university, a professor. At last, I found my knowledgeable expert!

"Sam, I need to discuss a question that has been part of my memory for several years."

"It's about the moon," I continued, "the astronaut's foot impression showed less than an inch of dust."

Sam agreed, his comment was, "The scientists at NASA expected to find over 40 feet of dust, the moon being billions of years old, they thought, cautiously including huge pods attached to legs of Columbia, the Lunar Lander, avoiding it's disappearance in a sea of dust."

"Maybe the moon dust floated to the valleys," Sam was prompt to add. "Later, the Lunar Rover also explored valleys, the lunar dust being of a similar depth," I was ready to point out.

The next statement by Sam confused me, that is, until recent years. Sam by now was visibly annoyed. "I can see that you are of the young recent earth theory, and I am not going there at my age," he stated.

My conversation with Sam before this was always friendly, stimulating, and interesting, enjoying his company. Our discussion ended abruptly this time leaving me disappointed and confused. Obviously, I am not able to have a constructive conversation with a university professor, at least not on this subject!

Why does a university professor refuse to accept new evidence or even discuss the possibility that the information passed down to him may be incorrect in lieu of recent evidence?

It is common knowledge that several tons of space dust land on our planet each day.[2] We are often told the moon is 4.5 billion (4,500,000,000) years old. In spite of this information, there is less than a depth of one inch of space dust on the moon![3] We know it did not blow away as there is no atmosphere on the moon, but there is gravity to keep the dust in place.

Is it time we questioned an original premise that the moon has been there for billions of years so knowledge may progress?

Time passed, but circumstances placed me in the same cottage over a weekend with Sam the Geologist. At that cottage, one of the guests and I decided to explore the vast forest across the lake, arriving there by boat. Our adventure included peering in a deserted cottage window, then a longer walk discovering a very deep chasm. The chasm was complete with a running water stream deep down at the bottom. It seemed so deep, this depression, the walls including a variety of rocks, all protruding, displaying a multitude of colors. An excellent discovery for our friend Sam to see, we concluded, he will tell us the geological history behind such an interesting discovery. Deciding to show Sam we proceeded back to our boat, then to the cottage to persuade Sam to see what we had discovered.

We traveled back across the lake with Sam once more while describing the strange depression found in the forest. We approached the location with caution not wanting to fall in. Sam looked around with interest, picking up several samples that lay beneath our feet. Then, he began to point to various colours and types of rock at the same time providing several names for rock types, a description that now escapes my memory. Each one of those rock types have an age of several "millions of years!" Again, those exact figures escape me, only remembering that they were indeed several million years old.

Satisfied, we resumed our trek back to the boat, while passing a series of blackish rocks that were most familiar to me. "These rocks," I remarked, "are similar to the ones I came across on my journey up Rougemont Mountain as a youth." Mentioning my observation to Sam he quickly identified the mountain connected with my youthful experience as part of the Monteregian Range of Eastern Canada, including a series of high places forming part of that region, not far from the banks of the St. Lawrence River.

"Many of the huge rocks are broken up," I continued to explain, "forming 'breadcrumb' like piles on every surface."

Sam's explanation came quickly, "Rocks do not last very long," as he persisted to defend his position. "They are subject to weather, continuous rain, heat of summer, then frost action in winter subjecting them to erosion." In addition, he went on to explain, "Rocks are subject to volcanic action, earth movement, including quakes, many actions of nature. Those rocks you describe in Rougemont are a result of ancient volcanic activity in that region of North America." How can some rocks be millions of years old while others last only a few years? I remembered the previous "moon dust" discussion with Sam and thought it wise to keep my surprised opinion to myself. It added to my confusion increasing my determination to discover the truth!

Am I coming to the correct conclusion when I learn that our University Students cannot question theories or opinions presented to them if they wish to graduate? Do our students need to repeat back teachings from theoretical ideas when other information is available?

That curiosity of mine led me to ancestral gravesites in New Hampshire to discover more evidence. The stones in graveyards rarely go back more than 500 years, granite that has eroded with age, only displaying partial lettering representing the life of one of my ancestors. That evidence of erosion and decay is apparent as archaeologists uncover the remains of massive buildings from once magnificent cities. These structures are a few thousand years' old, now partially buried pieces of rubble. The evidence all around tells us that rocks regardless of their type or condition do not last more than a few thousand years.

Why do we hear that some rocks last millions or billions of years when this is contrary to the rock cycle[4]? A rock is 300 million (300,000,000) years old, 100 thousand times older than archaeological discoveries, but it still exists today!

This is most confusing. Someone needs to show me how some rocks, the same type of rock, exist for several "millions of years," others return to dust in hundreds or thousands of years? Is there a major inconsistency here with evidence when one is claiming "millions," for rocks which is what we are constantly being told?

I once had a conversation with a brilliant young man, he may have some answers for me, satisfy my curiosity, someone to admire, a young man dedicated to scientific research. I know such a person, receiving high marks at university, in any subject, now a marine biologist, university professor, complete with certificate on wall proclaiming him, Doctor. He collects beetles and butterflies, placing them neatly in rows on trays, hundreds of them, all sizes and colours, carefully dissected, pinned with official Latin name inscribed beside them, labelled.

Does such a brilliant student believe in a Creator, a God, to account for all those beetles and butterflies, or did they just evolve over eons of time, an act of chance? This called for a direct question.

"Do you believe in God," I asked.

His answer came back just as direct.

"Never thought much about it," he said.

It seems that some questions linger in the mind; such was the case with this young student. I found him staring at me, more than a year later, with a stare suggesting some question or statement was there. "I believe there is a God," he said, as he continued to stare at me.

I was stunned at first. "Now what brought you to that conclusion," I asked.

He then explained how he had dissected hundreds of different beetles, similar, but none with the same features. "They had to be created at some beginning of time for a special purpose; namely, to consume all types of garbage, returning this consumed material to its original state, completing the lifecycle," he carefully explained to me.

I was not ready to accept this, returning his stated analysis with what I thought was a brilliant reply. "That would prove that these beetles evolved, one after another, over time," was my objection.

"No, that is not how evolution works," he explained. "These beetles were created just as they appear, at the beginning, a wide variety, complete with internal organs, a planned design ready to digest diverse garbage material, ensuring a continuing life cycle on earth."

Did I finally have some answers?

"May I offer you a piece of advice," was my next comment. Advice, I was preparing to give it, where or when had I heard it, maybe just a rumour, perhaps unhelpful gossip. I confess, this was a probe on my part, repeating what I heard somewhere, seeing if mine was prejudiced opinion.

This advice, I guess, was too long in coming for this student.

"Well, what is your advice," he inquired.

"It is my understanding that you are studying to obtain a doctorate," I commented.

"Yes that is correct," he quickly replied.

"This doctorate you shall obtain upon the advice of your professors?" I asked, continuing with a question.

"Yes," he said. "They have to all agree."

Finally, I issued the words, "If they find out that you believe in God, you will not receive that doctorate. I advise that you keep that discovery until after you receive the title."

The next set of words coming from this brilliant student shocked me, these words not soon forgotten. "Oh, I know that," he quickly admitted.

That strong statement made me think hard; knowledge being stifled, more recent information ignored; progress not encouraged!

I observed some of this young man's discoveries in the marine biology field. He described his many discoveries with papers accepted or rejected by his peers. He related how one of his discoveries is accepted, with later research he found to be false.

"So it will be rejected?" I was quick to ask.

"No," was his answer, "It will not change."

"Will all subsequent research be based on that false premise?" I asked.

"It will not be changed because it has been accepted," he explained.

I wonder how a young student resists his peers. What has happened to education?

What will happen to this brilliant student if he chooses to disclose what he has discovered? Will honest discovery and observation stifle his future progress if it contradicts the belief of the "status quo?" Does yet another new generation of students receive the same knowledge that is possibly out dated? I believe that rather than teaching our University students outdated information we should be encouraging them to think for themselves.

I was determined to find out more; my next discovery, adding to my knowledge, came from Oxford, England. Oxford; is a University town, very interesting with its old buildings and history wherever you look. There must be treasures to find behind these tall walls and rows of buildings, something for me to discover. An old church came into view, complete with steeple, requiring a second, and third look up, to see the top of this tall imposing structure. This church, beautiful inside, stained-glass windows, gold ceremonial vessels on table placed at end of long aisle stretching between rows of pews, on either side. Today, I was not here to worship, rather to find the entrance to stairs leading to the top of that impressive steeple to see all of Oxford from that vantage point.

An attendant kindly showed the location of the stairs with a stern warning. "They are very steep, with no handrail, please proceed cautiously."

This piece of advice did not stop my journey upward. The stairs were indeed steep and narrow with deep depressions in the centre of each stone step, evidence of other journeys in the past. Round and round, and upward I climbed, finally reaching the observation deck with a view that was well worth the climb. My first scene as I looked down was chimney pots, very distinctive, rising from rooftops, Gargoyles on the corners of every building in sight. I had a different view of Oxford here, something not seen from the street, patches of green everywhere, only accessible through narrow laneways. What an impressive sight of Oxford! Pubs, everywhere you look, typical

English meeting places. I planned to visit one of these pubs, talk to some students.

I found a small pub, first passing through a beautiful park to arrive there. I needed to bend low to enter; this old door with narrow entrance was evidence of a time when humans were much shorter and smaller. Three students at a table caught my attention, offering to buy them a "pint" for light conversation.

"I am from Canada," I explained. "One of my ancestors attended Oxford, graduating in the late 1500s." That certainly caught their interest.

"I read how my ancestor had many debates with the professors, debating, part of the educational agenda, back then," I continued to explain. "Do you still debate with your professors?"

"No," was the quick reply. "We listen to their lectures, repeat back what we hear; this is the only way to get credits from any university, at exam time."

My visit finally ended with these three students, but I came to an obvious conclusion. Listen to lectures from professors, no disagreement or debate, but only repeat back what you heard, accurately, at examination time, to receive the highest grade possible!

This is how you graduate from Oxford? Each new graduating student repeats what one's teacher learned without thought or updated information! Is education in Britain no different from higher levels of learning in Canada? Do University Professors not insist that their students think for themselves? Is it an objective of universities to get their students to think, not simply follow? Do we need "Free Thinkers[5]" equipped to handle the challenges that the future will present?

These are some of the red flags discovered by me! What have you discovered, and how will you handle them?

Therefore, it seems that we cling to theories long since outdated at the expense of progress. I now understand my Geologist friend when he aged those rocks, repeating back what he was taught, unwilling to question or change his beliefs. Are we continuing to promote outdated ideas?

In the following chapters, I explain the source of these theories and ideas with a description of personal experience challenging those claims. This information is for us to consider!

* Not his real name

References:-

1. Apollo 11 This page was last modified on 24 July 2014 at 22:15 24 July 2014/ <http://en.wikipedia.org/wiki/Apollo_11>

2. Cosmic Dust, This page was last modified on 23 July 2014 at 16:00 24 July 2014/ <http://en.wikipedia.org/wiki/Cosmic_dust>

3. Ibid.

4. Rock Cycle This page was last modified on 11 August 2014 at 17:29 19August2014/ <http://en.wikipedia.org/wiki/Rock_cycle>

5. Freethought This page was last modified on 21 August 2014 at 21:16. 01September2014 / <http://en.wikipedia.org/wiki/Freethought>

Chapter Two

Birth of Evolution Theories

How did we get this way? "A little bit at a time," was my reply.

I had just completed the analysis of each department of a company prior to implementation of a new integrated computer system. The chief executive of this manufacturing company was surprised with the quantity of reports laboriously produced by hand each month. Information in the form of written or printed reports requested at some time, still produced continuously each month, and probably no longer needed. My comment promised that most of these documents would no longer be required with the proposed computer system, complete with a visual screen, in each department.

This is common today, but I was introducing the first idea of instant access, a very radical idea only a few years ago. Prior to this, all data went to a separate Data Processing Department; data keyed in, and then printed reports produced. This is "batch processing," with little regard to the production of timely information. The system proposed demanded a unique computer, special operating system, and interactive program language, very unusual for computer systems sold only a few years ago for the business world. This trend continues. For a bank, I was part of the development to replace the computer-printed bankbook with access to the bank account on one's home computer.

All these systems required a logical set of conditions coming together to produce successful results. It required a person with detailed knowledge of business systems, someone to convert that knowledge into design for each program, then hundreds of programs, each working together.

"Get prepared to grow; your level of customer service will soon increase resulting in greater demand for your products," I explained.

I had trouble containing my pride as I often observed the result of computer systems previously installed. The business world demands that you grow. Business must adopt new but proven methods, continuing to be profitable or replaced by competition! Now I was preparing to analyse a very different set of conditions, which turned out to be the established Theory of Evolution.

My curiosity had peaked when I heard the strange stories that I describe in the previous chapter called "Red Flags," propelling me to find answers! Dare I use my skills to research an area of science not familiar to me? I had successfully installed a computer in nearly ever type of company, had to learn the workings of each different type of enterprise, never ignoring the smallest detail, nothing left to chance or the results would be a disaster. Was I qualified to attempt such a task when it applied to science? I remembered what someone once told me; a person can become an expert in any field if he chooses to put his mind to it.

12

That indeed is what I had done with computers and business systems most successfully, now I was prepared to do the same with Theoretical Science. I proceed to gather a wealth of information, many times questioning what I had found, facts discovered that were contrary to much popular belief!

I started this chapter recalling my experience with the manufacturing company mentioned. How did we get this way? I did this for several reasons but the first was that in my research I soon discovered that the Theory of Evolution developed a little bit at a time rather than all at once. It is not one theory but several, developed over a considerable period. Much to my surprise, each one of these theories was not based on Empirical Science, but rather from ideas expressed by men coming from "circles of influence" in the late 18th and early 19th centuries, nearly 200 years ago. Many other influential persons including some from the established Christian church then promoted the ideas of "evolution." This practice of "promotion" without scientific evidence continues today.

How did this happen?

What were the causes or conditions behind such an action? The following may present some ideas that provide answers. I suggest that the scientific community was annoyed with the established religious system before the Theory of Evolution arrived.

This is my reasoning.

In 1543, Copernicus explained that the earth was not at the geometric center of the solar system. In 1610, Galileo with the use of a telescope provided support for the idea called the Copernican Revolution. This statement that planet earth was not central to our solar system was in defiance of church doctrine, their incorrect interpretation of the Biblical account. For this radical scientific discovery, Galileo receives house arrest; such was the power of this religious system.[1] This indeed was a mild sentence!

To defy the truth as defined by that religious control system usually resulted in someone becoming a social outcast, ostracised, burned at the stake, tortured, imprisoned, or any other means necessary to silence such contrary opinions.[2] It became obvious that planet earth was not central for a very logical reason that I explain in another chapter!

In lieu of these recent scientific truths, such as Galileo's discovery, it seemed that the Biblical narrative of Jewish historical record was probably ancient superstition or myth. The scientific community readily adopted the idea of ignoring any supernatural event as described in the Jewish historical record. Prior to this most scientists accepted belief in a Creator with supernatural creative processes. This belief provided a balanced out look in their research, coming up with some of the most important scientific discoveries.

However, the stage was now set for scientists to present their ideas based only on theory. A person of influence may present a theory which was then accepted. The following outline of "Evolution Theories", the contributing sources, complete with influential support, is for your consideration. The contributors are extensive with my list certainly not complete, only meant to provide some basic input. Ideas presented by these influential circles produced the "Age of Enlightenment" also called the "Age of Reason."[3] But first, let me explain some events that I believe influenced this new thinking.

Prior to this "Age of Enlightenment," Europe was under the influence of a religious system established by the Roman Catholic Church called the Holy Roman Empire otherwise known as the Reich. Europe saw the choosing of Emperors for centuries providing a military force.

Holland fought with Spain for 80 years under the domination of this religious system obtaining freedom in 1581, defeating Philip II, then setting up the Dutch Republic. Philip II king of Spain was the son of Charles I head of the Holy Roman Empire, the 1st Reich.[4]

In England, Queen Mary joined with Philip II of Spain in marriage. This couple ruled the United Kingdom from 1553 to 1558. Queen Mary's father Henry VIII separated England from the Catholic Church to establish the Anglican Church of England. Once on the throne Queen Mary is determined to return England to the religious rule of the Roman Catholic Church. This queen adopted a reign of terror, with hundreds of people burned at the stake if they did not comply, including the Anglican Archbishop of Canterbury, Thomas Cranmer.[5] Others fled the country!

In 1558, after Queen Mary's death Roman Catholicism is removed by her younger half-sister and successor Elizabeth I. This new Queen fought the Spanish Armada led by Philip II of Spain defeating it in 1588.[6]

Freedom spread from Holland and England then eventually to the rest of Europe ushering in the Age of Enlightenment or Age of Reason. Free thinkers arose no longer subjected to the domination of a repressive religious belief system. The excesses of religion gave rise to rebellion against it. This was the beginning of Naturalism. This is a belief in nature only for changes in the world. There is no regard to history or supernatural forces. This also gave rise to Deism, a belief in God but rejection of established religion. These beliefs soon grasped by Atheism, no Creative God with statements contained in the

Humanist Manifesto. This also gave rise to many other isms of later centuries. Many founders of a new nation, the United States of America, shared deism. A democracy was born with the freedom principle forming part of their Constitution, including freedom from domination of their government by any religious system, often referred to as "Separation of Church and State."

A new "scientific belief system" in Naturalism is born. In addition, the "Age of Reason" is very selective only allowing a belief in Naturalism with total rejection of any historical records of a beginning including the Bible. The established academic community was now in a position to impose their beliefs in Naturalism as opposed to Creation spreading this ideology around the world.

I now had the answer to a puzzling question explained in a previous chapter of why a brilliant student would not receive a doctorate if he expressed a belief in any creative process rather than selection by nature, which is the doctrine of Naturalism. Influential men could point to events of the recent past to establish a new Worldview. The scientific community now moved from empirical scientific evidence to a belief in the imaginings of influential men.

The following is a sample of those mainly responsible for these new theories. I provide a summary here with extensive detail obtained over the internet using any search engine.

Baruch Spinoza (1632 - 1677)[7] was a Jewish-Dutch philosopher mainly responsible for laying the groundwork for the Age of Enlightenment. He came out of freed Holland. This movement was the beginning of Biblical criticism viewing the Bible as human rather than spiritual or of supernatural origin.

James Hutton (1726 – 1797)[8] was a Scottish Geologist known for introducing Plutonic Geology and Deep Time. He questioned the then current belief that Noah's flood laid down the Sedimentary Rock Strata known as Catastrophism. He introduced the ideas that "millions of years" laid down these layers of sediment. He laid the foundation for Uniformitarianism, which simply means "uniform action" over time, ignoring historical data written by the Jews as well as many other ancient peoples groups. These concepts and theoretical ideas were included in his publication called "Theory of the Earth (1785)."

John Playfair (1748-1819)[9] was a Scientist, Mathematician, and Professor of Natural History. He wrote a book in 1802, a celebrated volume, entitled "Illustrations of the Huttonian Theory of the Earth." This volume summarized and promoted the works of James Hutton later taken up by Charles Lyell. He was highly distinguished in the academic community.

Sir Charles Lyell, 1ˢᵗ Baronet (1797-1875)[10] was born in Kinnordy, Scotland. He became a British Lawyer and later foremost Geologist. He is a major contributor to many theories also promoting Hutton's ideas. He introduced the Geological column complete with imaginary ages in millions of years with Greek mythological names, the Eocene Epoch[11] being one of them. He promoted Hutton's idea of Uniformitarianism, a simpler term "The Present is the Key to the Past" idea came from the "Age of Enlightenment." Lyell's book, called Principles of Geology, (1830-1833) is an impressive three volume work. James Hutton and Charles Lyell considered the Fathers of Modern Geology, their opinions adopted by most Geologists today without question. Charles Lyell received many distinguishing honours from the scientific community including the prestigious Copley medal. Westminster Abbey is his burial place usually reserved for royalty, requiring the British monarch's approval, and adding to his influence.

Jean-Baptiste Lamarck (1744-1829)[12] was a French soldier, Naturalist, Academic, and an early proponent of the idea that Evolution occurred and proceeded in accordance with Natural Laws. He was a celebrated soldier, highly accomplished also recognized in Academic circles. His contribution to the evolution theories stated that complex chemical forces drove organisms up a ladder of complexity. Environmental forces then adapted to local environments making them different from other organisms. Charles Darwin promoted the idea of Micro to Macroevolution as his opinion for the birth of new species.

Very Rev. Dr. William Buckland (1784-1856)[13] was Dean of Westminster in the Anglican Church but also an English Geologist and Paleontologist. He wrote the first full account of the Fossil Dinosaur, which he called Megalosaurus. He published a book called "Connection of Geology with Religion explained in 1820. This book tries to justify the new science of geology with the biblical accounts of creation and Noah's Flood. He developed a new hypothesis that the word *beginning* in Genesis meant an undefined period between the origin of the earth and the creation of its current inhabitants, during which a long series of extinctions and successive creation of new kinds of plants and animals had occurred. This was his attempt to combine evolution with creation, the modern version of this commonly called Theistic Evolution. By 1840, he was actively promoting evidence of major Glaciations along with a new generation of Geologists such as Charles Lyell and Louis Agassiz. This provided a very strong influence for a belief in Evolution from a high official of the Church of England.

William Whewell (1794 – 1866)[14] was an Anglican priest born in Lancaster, England. He excelled in scientific disciplines providing many contributions to science. His best-known works are two books called "History of the Inductive Sciences (1837)" and "The Philosophy of the Inductive Sciences, Founded upon their history (1840)." He was a strong proponent of Inductive Reasoning[15] which the Internet reference explains as follows:

*"**Inductive reasoning**[16] (as opposed to deductive reasoning[17]) is reasoning in which the premises seek to supply strong evidence for (not absolute proof of) the truth of the conclusion. While the conclusion of a deductive argument is supposed to be certain, the truth of the conclusion of an inductive argument is supposed to be probable, based upon the evidence given."*

Scientists arrive at evolution theories by using this Inductive Reasoning[18] method.

William Whewell provided many new terms for discoveries including the word *science, physicist, Catastrophism, Uniformitarianism* and many others. In addition he received several honours for his work such as the Royal Medal (1837). He was a close friend and associate of Charles Darwin.

Charles Darwin (1809-1882)[19] was an English Naturalist and Geologist. He was a major contributor to the theory of evolution but not the only one, going the next step from the opinions of colleagues. He was a close friend of Charles Lyell influenced by taking a copy of his book "Principles of Geology" on his five-year sailing voyage on "HMS Beagle." Darwin was excited with Charles Lyell's theories then expanded on them. He proposed that all species of life descended from common ancestors. His book written in 1859 called "On the Origin of

Species by Natural Selection" covers these new ideas. "The Preservation of Favoured Races in the Struggle for Life" is the underlying title. Herbert Spencer, an English philosopher coined the phrase "Survival of the Fittest" as a simpler alternative description to Darwin's Natural Selection idea. A major ceremonial funeral honours Darwin with subsequent burial in Westminster Abbey. Darwin often described as the most influential figure in human history. Darwin's ideas influenced the world. Germany adopted Darwin's ideas with a prompt translation into German. This theory was instrumental in greatly influencing the academic community of that nation.

Thomas Henry Huxley (1825 - 1895)[20] was an English biologist known as Darwin's Bulldog. He was very influential in promoting evolution and instrumental in developing scientific education in Britain and elsewhere, against the extreme versions of religious tradition. In his 1863 book called "Evidence as to Man's Place in Nature," he states there is evidence for the evolution of man and apes from a common ancestor. He disagreed with Louis Pasteur's Law of Biogenesis that life only comes from life. He coined the word Abiogenist instead, an assumption that life arose from nonlife billions of years ago in line with evolution that he was promoting.

George Henri Joseph Edouard Lemaitre (1894–1966)[21] was a Belgian priest, astronomer and Professor of Physics at the Catholic University of Louvain. He proposed the theory of the expansion of the universe known as the "Big Bang Theory" as the origin of the universe. His theory does not state the kind of material, why it exploded, or who caused the explosion. This certainly seems like a bizarre proposal coming from a priest with knowledge of the Genesis creation story. Lemaire often receives the title "father of modern cosmology."

Science arrived here a little bit at a time. Many of the early pioneers in the development of science worked within a framework that also considered historical accounts. They saw in the geological record a testimony to the truth of the book of Genesis; including other records regarding Creation and a Global flood. However, in the late 18th and early 19th centuries, several assumptions began to control geological thought and the age of imaginary theoretical ideas began.

Therefore, I question Hutton's idea that sedimentary layers took "millions of years" to form with no erosion between them. This seems to be the beginning of many theories briefly explained in the following chapter.

For example, I observe layers of snow formed during any Canadian winter. One can observe layers of ice and snow deposited in a very short period, less than one month. Small deposited layers, the result of a series of snow fall. Then with warm spells during that time erosion develops between them. History also records a different reason for these sedimentary layers found around the world. They could easily be the actions of a series of flood deposits during a Global flood rather than very long ages with no erosion between the layers.

Is a decision by an influential few, years ago, to ignore any historical record one of the serious omissions for scientific research? These imaginary ideas, from a few influential men years ago, not based on any empirical evidence[22], are still used today without question and in most cases enforced on us! I do not limit my research to opinion, choosing to combine scientific knowledge with historical records coming up with some interesting results for you to ponder!

References:-

1. Galileo Galilei, This page was last modified on 23 July 2014 at 07:18 24 July 2014/ <http://en.wikipedia.org/wiki/Galileo_galilei>

2. Inquisition, This page was last modified on 24 July 2014 at 16:33. 26July2014/ <http://en.wikipedia.org/wiki/Inquisition>

3. Age of Enlightenment, This page was last modified on 23 July 2014 at 01:34 24 July 2014/ <http://en.wikipedia.org/wiki/Age_of_Enlightenment>

4. Reich, This page was last modified on 23 July 2014 at 01:45 24 July 2014/ <http://en.wikipedia.org/wiki/Reich>

5. Thomas Cranmer, This page was last modified on 26 June 2014 at 18:47 24 July 2014/ <http://en.wikipedia.org/wiki/Thomas_Cranmer>

6. Mary I of England, This page was last modified on 14 July 2014 at 20:23 24 July 2014/ <http://en.wikipedia.org/wiki/Mary_I_of_England>

7. Baruch Spinoza, This page was last modified on 25 July 2014 at 00:18 24 July 2014/ <http://en.wikipedia.org/wiki/Baruch_Spinoza>

8. James Hutton, This page was last modified on 17 July 2014 at 15:53 24 July 2014/ <http://en.wikipedia.org/wiki/James_Hutton>

9. John Playfair, This page was last modified on 15 July 2014 at 11:12 24 July 2014/ <http://en.wikipedia.org/wiki/John_Playfair>

10. Charles Lyell, This page was last modified on 22 July 2014 at 23:3 24 July 2014/ <http://en.wikipedia.org/wiki/Charles_Lyell>

11. Eocene Epoch World of Earth Science. 2003 05september2014/ <http://www.encyclopedia.com/topic/Eocene_epoch.aspx>

12. Jean-Baptiste Lamarck, This page was last modified on 16 June 2014 at 12:19. 24 July 2014/ <http://en.wikipedia.org/wiki/Jean-Baptiste_Lamarck>

13. William Buckland, This page was last modified on 15 June 2014 at 07:46 24 July 2014/ <http://en.wikipedia.org/wiki/William_Buckland>

14. William Whewell This page was last modified on 23 August 2014 at 09:05 25August2014/ <http://en.wikipedia.org/wiki/William_Whewell>

15. Inductive Reasoning This page was last modified on 17 August 2014 at 17:00. 25August2014/ <http://en.wikipedia.org/wiki/Inductive_reasoning>

16. Ibid.

17. Deductive Reasoning This page was last modified on 4 August 2014 at 15:29. 25August2014/ <http://en.wikipedia.org/wiki/Deductive_reasoning>

18. Ibid.

19. Charles Darwin, This page was last modified on 13 July 2014 at 14:52. 24 July 2014/ <http://en.wikipedia.org/wiki/Charles_Darwin>

20. Thomas Henry Huxley, This page was last modified on 21 July 2014 at 08:19 24 July 2014/ <http://en.wikipedia.org/wiki/Thomas_Henry_Huxley>

21. George Lemaitre, This page was last modified on 17 July 2014 at 16:26, 24 July 2014/ <http://en.wikipedia.org/wiki/Georges_Lema%C3%AEtre>

22. Empirical Evidence This page was last modified on 20 February 2015, at 09:29, 24 February 2015/ <http://en.wikipedia.org/wiki/Empirical_evidence>

Chapter Three

To Assume

Having no answer to my question, looking straight at me, he said, "But, I will find out." My thoughts then turned to experience; I've heard that one before.

It was science class, I remember that high school teacher very well. I was attending High School in the Eastern Townships of Québec, science and chemistry were my favourite subjects, with history a close second choice. Sometimes experience with teachers before meant promises, but little result. Was this teacher different? Well, maybe, at least he was willing to admit he did not have an answer.

The next science class, first thing, that teacher looked straight at me. "I have found the answer to your question," he said, then proceeded to explain the result of his research.

That experience had a profound affect on me, often remembering it later in life. The question is not important, I forget what it was now, but I never forgot the teacher. My career in business, as a consultant, I acknowledge that I do not have all the answers, admitting that I need more information, then quick to return a helpful reply. Honesty in business and life is recognized as a desirable asset.

Science became somewhat of an obsession with me after that incident. With a keen interest applied to every experiment, I learned much in the lab, completing Provincial exams with confidence that year, even believing I deserved a perfect mark.

The science exam for me this time seemed easy, one question being asked, what happens when you plunge a match, (it was actually a taper) into a jar that contained oxygen and hydrogen gases? It was worth eight marks.

That lab experiment performed by me many times had dramatic results. It was easy for me to explain that hydrogen and oxygen gasses convert to water, H_2O, with the application of a match. An impressive "little bang" resulting with a small amount of water found at the base of the container used for the experiment. I anxiously waited and wondered, how well did I do on that Provincial exam? Dare I believe a perfect mark! The 92 mark received was certainly very good, but which question was wrong?

Inquiry from the Provincial Examiner was not easy to find, but I persisted. That simple question about the H_2O experiment cost me eight marks. What happens when you plunge a match into a jar that contains hydrogen and oxygen gas?

"The answer is that nothing happens," the examiner promptly told me. He insisted that the question was very specific.

The match is not "lit!"

"This is trickery," I exclaimed, but he insisted that my answer was wrong. I was furious! How could a Provincial Examiner make such a mistake?

It took me some time to understand that indeed this "trickery" was for my education. In performing an experiment or delving into a particular situation, one must never disregard any factor, or make assumptions that may not be true. That one question, costing only eight marks was more beneficial to me in my later career than all the other questions combined. That lesson remembered becoming a computer consultant later in my career in business. In researching the exact requirements of several diverse businesses prior to installing a new computer system, never assume.

The original idea of Evolution assumed that no catastrophic event ever took place on this globe for "millions of years." "The present is the key to the past" was the expression of the day. These "millions of years" later changed to "billions of years," a decision that a much longer period was essential. A new belief system was born, a belief in the selective actions of nature only, ignoring all other considerations. It was summarily "assumed" that ancient Jewish historical writings were incorrect or just simple stories. No "catastrophic event" ever occurred such as a Global flood that would drastically alter our planet on both land and sea. This new proposal is finally adopted. Indeed, Jewish historical records as well as similar records of a flood from other races not considered. This is the accepted assumption by most of science even today, seldom questioned.

Indeed, not just one, but several assumptions form the basis of the "Evolution Theories", our new scientific belief system, used as the basis for all research. Some of these theories come out of the "Age of Enlightenment" of the 18th and 19th century.

Theories still accepted today as truth follow:

Deep Time Theory[1], our planet is millions of years old, later changed to billions. This idea proposed that all developed, including complex life forms from a single cell or molecule over long time.

Geologic Time Scale[2], a sub-division of the Deep Time Theory[3] into assumed ages complete with names for each, before discovery of any radiometric dating system.

Ice Ages Theory[4], (No Catastrophism[5]) a reported flood, deluge of water, is just local not Global. A new Worldview allowing us to imagine geological structures on earth created during eons of time, with a series of ice ages (Glaciations[6]), rather than the actions of massive amounts of water.

Uniformitarianism[7] ***Geology***, a complex description that simply means Uniform Action, that "the present is the key to the past." This new idea denies any violent actions even over billions of years, choosing to ignore the laws of nature that all material returns to dust over a very short period.

Micro to Macro, that all material evolves, becoming more complex. A condition contrary to the Law of Entropy[8] that states that all goes from complex to simple, from order to disorder. A condition of order to disorder that we witness daily.

Natural Selection[9], all plants and animals, came about by nature alone, over time. This new idea changed the Law of Biogenesis[10] discovered by Louis Pasteur that all "Life comes from Life."

These ideas form a very necessary part of the overall "Evolution Theories", assumptions started over 200 years ago, but never proven. These unproven ideas then progressed to form an extensive belief system. Some of these beliefs listed below:

Jewish History, of a "beginning" considered simple stories, ancient myth, or poetry.

No Intelligent Creation, all kinds of plants and animals just evolved by chance of nature over billions of years.

Secular Humanism[11] Beliefs, humans will determine morality by only using science and philosophy.

No Supernatural Events, only natural events considered.

Natural Selection[12], a belief that no intelligent design is necessary, all selected by the chance of nature.

Survival of the Fittest[13], that a human is not special, but subject to constant struggle for survival.

Master Races[14], some races evolved to a higher evolutionary level being superior to other races.

Complexity, through a series of deaths over time, all life forms became more complex, contrary to a scientific law stating that all goes from "complex to simple."

Big Bang[15], the universe created itself including everything in it by an explosion happening billions of years ago, some unspecified "force" involved initially.

We do not know what happened millions or billions of years ago, we have to assume, not having a definite scientific way of discovery, over such a very long period. In fact, evolution, being a set of theories, starts by assuming that the age of our beginning is in millions or billions of years. The theories progress from there, adding several more assumptions to form the hypothesis. Recent knowledge gained over the last 150 years, seldom considered.

Why do we never consider recent knowledge?

To answer that question one must understand how scientists go about doing their research and obtaining knowledge. Conclusions result from the "Scientific Method"[16] of research briefly explained and paraphrased as follows,

Question *Using the scientific method, everything starts with a question. Someone has a strong urge to answer the question. It is sort of a personality trait of a scientist.*

Answers *Then, there comes the answer. You have to be able to offer some ideas that confirm every answer that you give. Others must be able to test your answers as well.*

Evidence *Then there must be as much evidence as possible obtained by you and other scientists to prove your answers.*

Final Statement *A highly recognized scientist or group of scientists must accept the conclusions and evidence for it to become a scientific fact. A premise or benchmark is established.*

Subsequent Research The *scientific community must accept the premise or benchmark, and then all other research in this area must conform and assume that the original premise is correct.*

A problem arises when new research proves the old premise incorrect, as I previously explained with the university professor and brilliant student studying for his doctorate. All subsequent research must conform to the original premise or premises even if they are incorrect!

Is there no way to change assumptions that are centuries old, found in error? It seems logical to me that if the original premise is incorrect then all subsequent research or theories based on those premises cannot be true.

However, the embarrassment increases as modern research and technology add more theories or uncover the error of outdated ideas and beliefs. We continue in our effort to prove old theories without consideration of recent knowledge or a detailed historical record of a beginning.

Therefore, to assume may satisfy the need for a naturalistic worldview, but do these theories provide a realistic viewpoint, conforming to the truth?

My research is not restricted to previous assumptions but based on personal experience that may provide a different worldview. These ideas came to me after considerable research, information readily available to all of us now in this the "information age."

References:-

1. Deep Time, This page was last modified on 7 June 2014 at 23:27.02September2014/<http://en.wikipedia.org/wiki/Deep_time>

2. Geological Time Scale This page was last modified on 2 September 2014 at 01:26. 02September2014/ <http://en.wikipedia.org/wiki/Geologic_time_scale>

3. Ibid.

4. Ice Age, This page was last modified on 21 August 2014 at 18:38. 02September2014/ <http://en.wikipedia.org/wiki/Ice_age>

5. Catastrophism This page was last modified on 18 August 2014 at 22:29 02September2014/ <http://en.wikipedia.org/wiki/Catastrophism>

6. Glacial Periods This page was last modified on 23 June 2014 at 01:41. 02September2014/ <http://en.wikipedia.org/wiki/Glacial_period>

7. Uniformitarianism This page was last modified on 3 August 2014 at 13:46. 02September2014/ <http://en.wikipedia.org/wiki/Uniformitarianism>

8. Entropy This page was last modified on 25 August 2014 at 04:01. 02September2014/ <http://en.wikipedia.org/wiki/Entropy>

9. Natural Selection This page was last modified on 2 September 2014 at 01:29. 02September2014/ <http://en.wikipedia.org/wiki/Natural_selection>

10. Biogenesis This page was last modified on 6 August 2014 at 10:53. 02September2014/ <http://en.wikipedia.org/wiki/Biogenesis>

11. Secular Humanism This page was last modified on 20 August 2014 at 06:15. 02September2014/ <http://en.wikipedia.org/wiki/Secular_humanism>

12. Ibid.

13. Survival of the Fittest, This page was last modified on 26 August 2014 at 23:21. 02September2014/ <http://en.wikipedia.org/wiki/Survival_of_the_fittest>

14. Master Races, This page was last modified on 30 August 2014 at 07:27. 02September2014/ <http://en.wikipedia.org/wiki/Master_race>

15. Big Bang This page was last modified on 2 September 2014 at 00:30. 02September2014/ <http://en.wikipedia.org/wiki/Big_Bang>

16. Scientific Method, This page was last modified on 14 July 2014 at 18:50 25July2014/ <http://en.wikipedia.org/wiki/Scientific_method>

Chapter Four

Deep Time Theory

"There is nothing so powerful as truth - and often nothing so strange" Daniel Webster.

From the top of that building, a laser beam stretched out into the distant horizon. Below our feet ran a straight silver line with zero (0°) marked on it, I had one foot in the West, and the other in the East. The Prime Meridian of the World, Longitude zero (0° 0'0) where East meets West, this being the home of Greenwich Mean Time *(GMT)* from which all world time zones are based.

My wife Gillian and I were visiting England, attending her cousin's wedding. While there, I was determined to travel to Greenwich, get answers from yet another reliable source, answers to a question so troubling for me.

Our boat journey down the Thames was historical, passing the Tower of London, Big Ben, and the Houses of Parliament. Sailing under Tower Bridge, we passed a series of wharfs to arrive at our destination, Greenwich. The long climb up the hill brought us to the famous Royal Observatory, and once there we enjoyed a spectacular view of London.

Arriving with certain questions, I wondered, whom could I ask?

There was an excellent presentation at the Planetarium. The big screen displayed planet Earth from space, complete with description and spectacular views of our Universe. The Astronomer present was most informative inviting us to ask questions at the conclusion of the presentation. Finally found, someone in the know, an astronomer, who better to answer my questions! Most people were too shy to ask, but not me!

Our Planet Earth, is the daily rotation slowing by an average of one second every 18 months? Without hesitation, surprising me, he answered my direct question.

"Yes, that is correct." "However, it will not stop rotating in our lifetime," he quickly added.

My next question was more direct. "I have learned that our Planet Earth is about 4.6 billion years old." "Is this true?"

"Yes, that is what we understand," was his reply.

My questioning continued! "If the earth is 4.6 billion years old, how fast was it rotating at the beginning for it to slow down so much?"

Gillian commented to me later, "I don't think the astronomer has ever been asked that question before." His answer was short. "I don't know," he said.

I now had confirmation from a knowledgeable person to a question so troubling for me, from an official at the Royal Observatory in Greenwich. Confirmation, our Planet Earth is indeed slowing down in its rotation, daily.

Prior to my visit to Greenwich, I learned from the International Earth Rotation Service and the United States Naval Observatory that the earth rotation slows as they explain, one second in 500 days, about 2 milliseconds per day. This is called

Leap Seconds, details accessible over the internet using any search engine. I had to be sure having doubts based on information from multiple but unofficial sources. In contrast, they stated 2 ms. in 100 years, or 5 ms. per year, or 1.5 ms. per century, others 2.2 seconds every 100,000 years, 0.0017 seconds for this century, all in spite of official contrary information.

Still questioning my discovery, I explained this information about Leap Seconds with planet deceleration during a recent local astronomical presentation. The suggestion given was that the "slowing" action was probably only recent and temporary.

Do I finally have an acceptable explanation, one that I could readily accept?

I accepted this explanation at first until my logic kicked in!

Our planet is the home of multiple life forms demanding exacting conditions. The moons gravitation causes tides twice a day. This action in time affects earth's rotation. These tides become essential for the health of our oceans, all life forms, also slowing the rotation of our planet. I was not easily deceived but had more questions to confirm my conclusions!

The first reference concerning Leap Seconds[1] comes from *Wikipedia Encyclopaedia* which provides a table of the slowing rotation of our planet since 1972, which comes to 35 seconds.

This site states there are plans to abolish the calculation of this slowing rotation of our planet. Has anyone considered what effect this changing balance in equilibrium has on the condition of our globe around the world?

The second reference comes from The *United States Naval Observatory* which specifically says an average of one second out of every 500 days. This is required to keep very accurate atomic clocks in coordinated locations all around the world in line with Earth's slowing rotation. These seconds are indeed called *Leap Seconds*.The following is a quote from that web site (tycho.usno.navy.mil/leapsec.html):

"Currently the Earth runs slow at roughly 2 milliseconds per day. After 500 days, the difference between the Earth rotation time and the atomic time would be 1 second. Instead of allowing this to happen, a leap second is inserted in the Atomic Clocks to bring the two times closer together[2]."

These findings also conform to certain universal Laws of Science:

***First Law of Thermodynamics*[3]** states that an outside force and energy is required to produce a new action in an enclosed system. This Globe, our Earth, is no exception to the rule. The First Law was required to produce a rotating action in the beginning!

***Second Law of Thermodynamics*[4], *the Entropic Principle*[5],** states that all objects in an enclosed system will eventually run down. Our Earth will eventually stop rotating due to several forces but mainly because of *Tidal Friction*[6] coming from the Moon.

We can easily determine the approximate length of time that this globe has been rotating since there are 86,400 seconds in a day. This is a figure arrived at by multiplying 24 hours in a day by 60 minutes in an hour by 60 seconds in a minute with a result of 86,400. With a slowing rotation of one second every 500 days, once every year and one half it would take 129,600 years for our planet to stop rotating, but logically much less.

37

This provides a computation in thousands rather than millions or billions of years for this planets existence at least when converted to support life forms! A constant spinning rate of approximately once every 24 hours is essential for an equal period of night and day producing a relatively constant temperature.

Therefore, with recent knowledge, do we need to review the "assumptions" presented to us during the Age of Enlightenment[7]? Faith in the wisdom of only a few influential men allowed us to accept very long ages for our living planet in the past. The historical record of Genesis states only a few thousand years now with confirmation from science.

However, recent scientific knowledge also suggests that our planet is 4.5 billion years old! Which age is correct? These two ages may be a conflict between science and history, easily explained by consulting historical records. Both ages are possible! I explain this age conflict in a subsequent chapter.

Technical Data: *

For the person who likes technical details, planet Earth is a marvel of perfection traveling 940 million kilometers (584 million miles) each year. A huge object 40,233 kilometers in circumference (25,000 miles), it is 12,756 kilometers in diameter (7,926 miles), complete with gravity, perfectly balanced, rotating at more than 1,609 kilometers per hour (1,000 miles per hour) at the equatorial circumference. The rotation of our planet maintains close to Perpetual Motion only losing about 1 second of rotational momentum every 500 days. This is beyond anything that we can imagine or believe from experiences here on earth. What a marvel of engineering perfection!

Additional Information available by using any Internet Search Engine:

Sidereal Time

Tidal Friction

Tidal Acceleration

Tidal Deceleration

Tidal Locking or Captured Rotation

Locking of the Larger Body ·

Newton's Law of Motion

Entropy

Empirical Evidence

Leap Seconds

* Figures are approximate

References:-

1. Leap Second - Wikipedia This page was last modified on 16 July 2014 at 07:24, 25 July 2014/ <http://en.wikipedia.org/wiki/Leap_second>

2. Leap Second - United States Navy "Time Service Dept., U.S. Naval Observatory, Washington, DC,"/ 25 July 2014 <http://tycho.usno.navy.mil/leapsec.html>

3. First Law of Thermodynamics, This page was last modified on 15 July 2014 at 00:00. 25 July 2014/ <http://en.wikipedia.org/wiki/First_law_of_thermodynamics>

4. Second Law of Thermodynamics, This page was last modified on 24 July 2014 at 08:28, 25 July 2014/ <http://en.wikipedia.org/wiki/Second_law_of_thermodynamics>

5. Entropic Principle, This page was last modified on 24 July 2014 at 11:18, 25 July 2014/ <http://en.wikipedia.org/wiki/Entropy>

6. Tidal Friction, This page was last modified on 10 July 2014 at 13:00, 25 July 2014/ <http://en.wikipedia.org/wiki/Tidal_friction>

7. Age of Enlightenment This page was last modified on 3 September 2014 at 23:09 04September2014/ <http://en.wikipedia.org/wiki/Age_of_Enlightenment>

Chapter Five

Global Flood or Ice Age Theories – Glaciations

I watched from the protection of a small shed, as three hours of heavy rain descended. This was a deluge, causing a small creek to overflow its banks; the water taking large pebbles, moving them, and then rolling them to another field. Water was rushing down the valley in the distance, as I watched, creating a miniature Grand Canyon. The successive movement of water left layers of dirt in its path as it rushed down to the bottom of the valley, muddy water reaching the otherwise small babbling brook below. I described this condition to a friend who proceeded to relate how he once observed trees and houses removed as he watched, floating them down stream during hurricane Hazel. Water was removing them, taking them along with the raging torrent, depositing them several miles away, downstream.

I live in an area about 40 miles north of mighty Niagara Falls in an old stone farmhouse complete with traditional red barn behind. This barn is no longer used as a home for a variety of animals, evidence of a past. A large shed sits behind the old house, and then a small creek flows between shed and barn. The small creek meanders its way eventually reaching Lake Ontario. It was on a Sunday afternoon during a three-hour period that water rose in that small creek creating a rushing river that finally entered the floor of the elevated shed. I was relieved when the water on the surface of the floor did not exceed two inches. Water descended the valley in waves as I watched this deluge of water in the distance. The small elevation on one side of this long valley contained a series of ponds, one quite large. As rain descended on this large pond surface, the water began to rise eventually bursting over the sides. The escaping water then found a weak spot on the shoulder creating a ditch rapidly emptying much of the contents of this large pond. A new volume of water descended the valley on its way to the otherwise small creek below.

The small creek returned to its placid state when the rain finally ended leaving evidence everywhere. Soil including rock, some quite large moved from one field to another. The surface of valley bottom changed where rainwater descended at first, then massive amounts from that large pond appeared. Top soil and sediment were removed from valley base creating a miniature canyon. The rushing water left deposits of soil and sediment creating pancake like layers before entering the creek below. Such was the scene after a short but heavy deluge of water on an otherwise sunny afternoon. All evidence soon disappeared as grass grew in the fields once more. The sides of that miniature canyon soon caved in filling the space visible only a few days before. The passage of time does not produce, but rather destroys evidence of even recent events.

I can only imagine the massive events, never again occurring, that took place during the Global flood resulting from rain falling for 40 days even during the darkness of night. This, over 900 hours longer than the short three-hour period that I witnessed! The associated change to our planet causing massive convulsions connected with such an event is indeed difficult for us to imagine. The evidence, however, is Global, very apparent for us to observe and discover, if we are willing.

Can production of the Grand Canyon be a deluge of water only a few thousand years ago? If this impressive feature, the Grand Canyon, were millions of years old, would it still be visible with no apparent erosion?

Here in Caledon, region of Peel, in the Province of Ontario, Canada, there are large pools of water everywhere. Round pools have no inlet or outlet of water to form or empty them. This is a mystery for me until the answer came as I carried sap in a pail from maple trees one spring day. Water is heavy, very heavy, if lying in one place, very deep, thus the weight on the soil below must be very great. The very heavy water created round craters depressing the soil or aggregates in my particular area. I mention this as one can imagine water at some time rushing down an area of Colorado producing the Grand Canyon in its wake. I have difficulty believing that the small Colorado River is capable of carving such a deep chasm also moving heavy sediment in its wake even considering millions of years. In addition, the sides of this famous canyon would collapse falling to the bottom over such a long period.

Is there evidence of large pools, lakes, forming above the present Grand Canyon, finally spilling their contents descending as a massive deluge?

Do we have questions needing answers?

The effects and evidence of a flood including massive upheaval are extensive happenings in the recent past, globally. I shall mention only a few but the most evident is what someone chose to call the Cambrian Explosion[1]. A major amount of fossils appeared suddenly contrary to accepted Evolutionary Theories. Theories that states it normally requires billions of years to evolve. There is evidence that fossils appeared in massive quantities globally apparently 500 million years ago! In 1840, Darwin discussed this as one of his main objections to Natural Selection, part of the "Evolution Theories."[2]

How did this massive diversification of complex organisms, fossils, arrive over such a short period? May I simply suggest that this is a massive graveyard rather than life evolving? Is this simply evidence of many diverse creatures trapped during a recent but sudden Global deluge?

The effects of massive water are apparent everywhere on our planet! I often observe very official plaques stating millions of years ago, placed in the ground for tourists to read. One attraction, close to where I live, describes the "Bad Lands[3]" of Cheltenham in Caledon, ripples of hard red soil that the official plaque states being there for 430 million years, result of an ancient seabed, supposedly during the "Ordovician age[4]." Boys and girls run on the hard red soil, not a good idea stated on the web site; their little shoes in time will destroy the mounds of soil, something those 430 million years of existence has not been able to do!

This "Bad Lands" location in Caledon is roughly 40 miles from mighty Niagara Falls, as the crow flies; the plaque there reads 12,000 years that those mighty Falls started flowing not 430 million years ago as stated on that plaque only 40 miles north. This figure of 12,000 years readily proven, the deep Niagara Gorge[5] forms a beginning point of massive amounts of fresh water, fast flowing, now halting at the present location of those famous falls. That flowing water over the falls still lengthens the Niagara Gorge!

I learned later that the basis for the age of this Gorge is on assumed average erosion because of a belief in "uniform action[6]" theory used to determine age. We also learn that this water comes from the last "Theoretical Ice Age[7]" forming the Great Lakes, largest freshwater deposit in the world.

The "ice age" theories need explanation to identify the source of many ideas. In 1742 Pierre Martel, an engineer and geographer, found that small glaciers in Europe had sometimes extended much further south. Many other scientists adopted this new idea.

"The distant past may have provided a series of very cold periods resulting in the cover of whole continents with ice sometimes up to 4 km thick (about 2.5 miles)."

The scientific community admits that "The cause of the ice ages is not fully understood." Several new theories are then proposed such as, "changes in Earth's atmosphere," "position of the continents," "fluctuations in ocean currents," "variations in Earth's orbit," "variations in the sun's energy output," "volcanism," and the ideas continue to be presented to explain the "ice age" phenomena. In lieu of these theories the effects of these "ice ages" are then considered. The resulting effect of these "theoretical ice ages" then creates more theories.

A reference in the Wikipedia Encyclopaedia indicates that the last glaciations period formed the Great Lakes of North

45

America. Glacial ice covered the Northern Hemisphere then retreated north gouging out the Great Lakes Basin[8] leaving all fresh water behind. This is quite a story!

The by product of this massive flow of water creates Drumlins[9]. You will find this smooth hill like formation in various parts of the world not just in the Northern Hemisphere. Drumlins are simply the droppings of fast flowing water that left deposits of sediment in their wake as the water flowed to a lower elevation. I observe the many formations in Caledon where I live and they look like waves in the ocean. This rush of water, dropping sediment in its wake eventually reached a lower elevation, Lake Ontario, part of the Great Lakes. This massive quantity of water finds an ideal location causing huge depressions creating the Great Lakes Basin[10]. This is similar to the relatively small round "basin like depressions" filled with water that I find where I live which is part of that Great Lakes Basin.

May I suggest that no theoretical ideas such as "ice ages" are required to explain the production of the Great Lakes! May I also suggest using written human records that describe in some detail a deluge of water descending from the firmament above only a few thousand years ago, causing a massive change to the surface features of our planet globally? Has anyone examined the effect of massive amounts of water flowing in every direction in search of a lower elevation?

Water is a powerful force, dissolving hard material, carrying with it volumes of sediment for miles. Sedimentary aggregates are found in quarries most prevalent in the Caledon area, with rocks that I discover are perfectly round in shape due to abrasive rolling action. The creation of lakes to the south east, in New

York State, are evidence of massive water action as well leaving very deep fresh water lakes, the Finger Lakes, as the water continued on its journey. On the other hand, was this the spilling action of Lake Ontario as rain continued to fall on its large surface?

My experience in boarding school provides evidence of the need for much water to produce even a small amount of ice. The boys took turns during a very cold winter night to run water from a large hose. We used a fire hydrant to produce ice for a hockey rink, lots of water to make ice up to a foot thick, during a very cold winter period, soon melting with spring weather. A better example may be ice cubes from a home refrigerator. Water is essential in producing ice!

I find that the reference previously mentioned from the Wikipedia Encyclopaedia explains that the water to produce the "ice ages," ice that is sometimes 3 to 4 km thick (2 to 2.5 miles), came from the oceans around the world. This additional theory then states that this caused the oceans to lower by 110 meters (360 feet). This is what it says in the section called "Effects of Glaciations[11]."

"During glaciations, water was taken from the oceans to form the ice at high latitudes, thus global sea level dropped by about 110 meters, exposing the continental shelves and forming land-bridges between land-masses for animals to migrate. During glaciations the melted ice-water returned to the oceans, causing sea level to rise[11]."

This is what I understand from that official statement:

During a very hot period huge volumes of water evaporate from the salty ocean surfaces. This evaporation lowers the oceans by as much as 110 meters (about 360 feet). Land areas then receive this large volume of fresh water. An extremely cold period then develops forming ice in the form of glaciers that cover most of the continents on earth. This ice can be 3 to 4 km thick (2 to 2.5 miles). A new very warm period develops melting the ice leaving fresh water on land in lakes around the world. We know that the Great Lakes Basin contains a major quantity of fresh water on the planet that arrived there only a few thousand years ago due to the receding Niagara Gorge. The oceans then return to their previous water level. However, the return of water in the oceans did not come from river run off since most of the water stayed in lake reservoirs around the world like the Great Lakes Basin!

I have a question. Where did the new supply of water come from to fill the oceans up to the previous level once more?

Then I have other questions. What heat source caused so much water to evaporate from the oceans? What action created such massive ice that covering most of the continents? What changed to create a new very warm period to melt the ice? What happened to the "uniform action" theory (Uniformitarianism[13]) that states "The present is the key to the past," what we see today is the way it has always been?

Just a few questions but you may have more!

May I suggest that there was no one there to record such a drastic event! I also question how many natural laws this theory would violate! In addition, where was the sun that did not melt the ice over even a short period regardless of such a long time, making that massive quantity of ice impossible to form? These are only two very simple objections to the Ice Age theory. Do we need a fresh look at reality rather than accepting ideas that do not seem to "hold water?"

My research also included ice fields around the South Pole, ice over 3 km thick (over a mile) covering a vast area in a location dryer than any dessert, receiving little or no snowfall each year. It seems logical that a major amount of water arrived at one time from somewhere to form the ice. Is this the same volume of fresh water descending from that broken canopy above producing so much ice at the South Pole? May I leave that question for you to consider?

My simple investigation seems to indicate a need for serious change in "age old" thinking! The North American continent paints a very different Geological picture than the areas used to develop the original theories centuries ago. Am I being arrogant to suggest any new ideas?

I mentioned my findings to a friend who proceeded to ask if I knew more than our scientists. I tried to explain to another friend that my research proved our planet is only thousands of years old. He did not wait for any explanation. He asked me if I also believed that planet Earth is flat. End of discussion!

Therefore, all happened "millions of years ago," ideas now taught coming from theories of the 18th century. We are now in the 21st century with so much more information available from every part of the world, daily. Are our theories just as valid as ones coming from the 18th century with their limited access to information? Is it time we considered new possibilities instead of relying on theories centuries old, having much better tools at our disposal? On the other hand, are we just trying to match recent evidence to old theories without thinking?

References:-

1. Cambrian Explosion This page was last modified on 20 July 2014 at 02:56, 01Aug2014/ <http://en.wikipedia.org/ wiki/Cambrian_explosion>

2. Ibid.

3. Cheltenham Badlands This page was last modified on 13 September 2013 at 15:28, 01Aug2014/ <http://en.wikipedia.org/wiki/Cheltenham_Badlands>

4. Ordovician Age This page was last modified on 20 July 2014 at 21:44, 01Aug2014/. <http://en.wikipedia.org/ wiki/Ordovician>

5. Niagara Gorge This page was last modified on 6 June 2014 at 05:52, 01Aug2014/. <http://en.wikipedia.org/ wiki/Niagara_Gorge>

6. Uniformitarianism This page was last modified on 3 August 2014 at 13:46. 04August2014/ <http://en.wikipedia.org/ wiki/Uniformitarianism>

7. Ice Age This page was last modified on 8 July 2014 at 03:07, 01Aug2014/. <http://en.wikipedia.org/wiki/Ice_age>

8. Great Lakes Basin This page was last modified on 28 April 2014 at 07:52, 01Aug2014/ < http://en.wikipedia.org/ wiki/Great_Lakes_Basin>

9. Drumlins This page was last modified on 10 July 2014 at 13:53, 01Aug2014/ <http://en.wikipedia.org/wiki/Drumlin>

10. Ibid.

11. Glaciations This page was last modified on 20 July 2014 at 13:05 04Aug2014/ <http://en.wikipedia.org/wiki/Timeline_of_glaciation>

12. Ibid

13. Ibid

Chapter Six

Uniform Action Theory (Uniformitarianism)

That painting is wrong! The mountain is not that high! Would an artist make such a mistake?

I was visiting the museum of history in Ottawa, Canada. This interesting museum contains many relics, treasured pieces of art. An accomplished painter toured with me as we came across that piece of art, now very old. The Monteregian Mountains featured in the painting, only a few miles from the banks of the St. Lawrence River, a scene from my hometown, very familiar to me.

This was a beautiful painting! We concluded that the artists view came from Montreal, somewhere on Mount Royal, the highest elevation on the island.

"The artist is looking into the distance past the St. Lawrence River to those Mountains," I mentioned to my friend. That Rougemont Mountain is the highest in the distance, a French word meaning Red Mountain, place of my birth. "Only one mountain in that painting rose higher than the others."

Turning to my artist friend, I exclaimed, "That Mountain in the middle of that picture is too high. Why would an artist distort the height of one particular elevation when I know for a fact that it is not that high?"

She quickly exclaimed, "The painter would not distort the height, it had to be that high when it was painted."

Never forgetting that incident I kept wondering how that mountain known so well could change in such a short length of time. That question solved several years later during a conversation with Cousin John Standish. That cousin now lives in my ancestral grandparents' home at the base of Rougemont Mountain.

An old biography provided the answer to that question found by my cousin, also explaining another mystery for me. "Why was I born in this small town with its high mountain?"

The answer, quite logical, better to see enemy approaching from an ideal vantage point. These enemies could then sail along those rivers to capture Montreal, then capital of Canada. My ancestors, Captain Sias Bachelder and Lieutenant Colonel Daniel Bachelder were in the then British army. St. Thomas graveyard in Rougemont is the burial site.

The enemy for a brief period coming from Thirteen Colonies to the south now called the United States part of the rebellion against the mother country.

"They are cutting down the big oaks on Rougemont Mountain," the old diary stated. "They are floated down the river to Montréal. Oaks, tall and strong sent down the local river to Montreal then shipped overseas to Britain. Tall masts from Canada destined for use on British sailing vessels," our relative continued to explain in her diary.

In 150 years or less that mountain lost its original height eroding to its present size with the removal of those tall oaks. This is human action to a well-known landmark, but the actions of nature are much more violent. That mountain eroded to its present height in less than 150 years! Change all around us is a part of our existence on this our living planet. Geology alone cannot explain such a change without history. Is such explanation the preserve of history?

To the artist that mountain was all that he saw in the distance. The artist may also conclude that the scene is constant, has and will always be the same. Knowledge provided a reaction to this change that I now choose to record. This knowledge provides a very different point of view!

However, a new worldview started many years ago, an older "catastrophic event" part of the belief then, successfully removed! A decision to change history, a belief still taught today, to disregard any reference to a Creator or creation by the scientific or academic community. Consider only nature, the natural world was the decision of their day. That decision caused the birth of that new worldview. Rather than any catastrophic event the idea of "uniform action[1]" (Uniformitarianism) is still believed and applied by research. These ideas were formulated by influential Naturalists in the late 18th century, started by James Hutton, refined by John Playfair then successfully promoted by Charles Lyell with his popular 1830's book *Principles of Geology*[2].

William Whewell also a friend of Charles Darwin named the term Uniformitarianism[3] This big word simply means "uniform action" over time. The word later described as "The present is the key to the past," choosing to disregard any Global flood with associated major changes to our planet. Uniform

Action over eons of time is the only consideration! That idea even denies a relatively simple change to Rougemont Mountain in less than 150 years!

James Hutton assumed that layers of sediment arrived there over long periods, each layer being millions of years old. However, a closer look would reveal a lack of decay or erosion between the layers, a common occurrence of natural forces over even a short period. In spite of this obvious contradiction, Charles Lyell was able to convince colleagues of the truth of this theory. He then provided theoretical ages[4] to each layer including mythical Greek names, these names still used today. For example, he named the Eocene[5] epoch in 1833, being one of them, after the Greek word Eos with an imaginary or theoretical age from 54 to 38 million years ago! This certainly was quite a display of influence for the scientific community ages ago giving us a good example of the power of influence in shaping our future!

James Hutton and Charles Lyell considered the fathers of modern geology changed a previously accepted worldview. Prior to this new theory, it was common knowledge that a catastrophic event changed our planet in the recent past. The deposits of sediment that we readily observe science accepted back then as the result of a Global flood as recorded by history. Since the introduction of this new theory there continues to be a concerted effort to prove the existence of long ages a most necessary part of evolution. Is this challenge a constant need to reconcile scientific theories with recorded history?

Science claims billions while Genesis states thousands of years for our existence.

This presents quite a dilemma! If we carefully examine the historical record can we determine that both are correct? A

55

subsequent section in this chapter provides a possible explanation for the reason behind the truth of both ages!

Then other opinions arise to add to the confusion. We hear that the book of Genesis needs proper interpretation or the record of a beginning in Genesis is simple story, mythology or even poetry. Some stories tell us that the Global flood story was just a local occurrence. Then, some say that the subject is not for the average thinker but requires a highly educated person to provide an interpretation for us. One story even tries to convince us of billions of years between days[6] in spite of a clear record of days in the book of Genesis that includes evening and morning each day!

Is all this an attempt to reconcile two opposing stories, creation versus evolution?

This attempt at reconciliation is clearly not possible as noted in the following brief comparison of these two opposing worldviews:

Creation Worldview; disobedience by the first two created humans at the beginning introduced an error in nature causing "death." The continuance of creation involves a relationship between male and female for humans, all living creatures, including plants. DNA instructions essential for continuing creation built into each "kind" by the Creator.

Evolution Worldview; a series of "deaths" created all we observe. This happened over eons, billions of years. This theory proposes small changes upwards in complexity guided by nature. Natural forces alone were choosing a series of "deaths" as a means to evolve everything from a single cell or molecule to the complexity of all nature, including humans.

These are two very different worldviews! In spite of this new evolution worldview, we do not observe anything evolving. We continue to search for transitional forms in the fossil record without success. This lack of supporting evidence often called the "missing links."[7] We do however continue to observe relationships everywhere in nature creating new offspring including humans.

So which is correct, the roughly 6000 years for the creation story in Genesis or the 4.5 billion years proposed by science? The following provides a possible answer to that dilemma!

A careful reading of Genesis, the first book of the Bible written by Moses, may provide that answer. It is wise to use an older version of the Bible such as the King James Version, or use a Jewish version such as "The Holy Scriptures according to the Masoretic Text" as I did, newer versions may add to the confusion.

The first sentences in those two versions say "heaven," singular, rather than "heavens," plural. This can be a very important difference, as I will explain. These versions continue to describe "heaven" to be a "firmament" above and surrounding the earth, now commonly called atmosphere, not the total universe as in "heavens." It then says the earth was unformed and void, planet earth then designed to support life. The story describes a very logical sequence of events in days required to prepare what I call the very old "dead planet" into one that supports all life forms a few thousand year ago.

The story continues, and it is significant that Genesis soon mentions a water-covered planet. SETI, (Search for Extra Terrestrial

Intelligence)[8] have used telescopes around the world for several years without success to discover any planet, anywhere, with an abundance of water similar to ours. Now we may understand how a universe including our then "dead planet" existed prior to conversion of planet earth into one that supports life.

How do scientists age material found on planet earth? They cannot age rock as it gathers a variety of material, becoming a compound. The ageing process is only possible using the "elements" found in the rock, created at some time in outer space. For example, uranium changes into lead over a very long period rather than just a few thousand years. This is what is most important to consider!

Those "elements" were present in planet earth before that Spirit converted an unformed planet into a living one estimated to occur about 6000 years ago according to biblical historical record! Further confusion arises when we assume that all other material including fossils in a rock are the same age as some metal "element" contained in it!

Do we also assume that any other material lying close by, such as dinosaur bones to be the same age as the metal "elements" in the rock?

The mystery continues, but is the story in Genesis acceptable as scientific fact? Does it comply with the known scientific model? The following provides an interesting observation;

A noted atheist and philosopher Herbert Spencer made an impressive discovery in the late 1800's. He stated the categories of the "knowable" and was then hailed for this discovery by the scientific community. He determined that everything that exists fits into one of five categories, "time, force, action, space, matter."[9]

The first sentence in the Bible, book of Genesis, says "In the beginning," that is time, "God" that's force, "created" that is action, "the heaven," that is space, "the earth," that is matter. Everything that Herbert Spencer discovered is found in the first verse of the Jewish historical record written thousands of years ago. These are certainly not the writings of an uneducated nomadic tribe as some authors choose to imply.

We come up with several theories associated with evolution. The most bizarre idea is a "Big Bang"[10] happening billions of years ago, creating everything. We have little means to discover what happened millions or billions of years ago but we have much evidence all around us now. The following is just a few examples of supporting evidence for very recent activity, difficult to deny:

Niagara Gorge, the starting point of the gorge created by mighty Niagara Falls is less than 10,000 years old. The action of fast moving water eroding the land to the present location producing a gorge, result of a daily release of massive amounts of water from the largest fresh water deposits in the world, cascading over those Falls. The Great Lakes releasing the water obviously created during a similar time as the Niagara Gorge.

River Deltas, created around the world, only a small deposits of sediment at the mouth (river deltas[11]) providing evidence of recent actions of water flow, not flowing for millions of years.

Polar Ice, this is a massive deposit of ice from freshwater at both north and south poles. The ice at the South Pole is more than 3 km thick (over a mile) in an extensive area dryer than any dessert. Is this more evidence of the arrival of massive quantities of fresh water. Is it possible to make ice without fresh water being present first?

Sedimentary Layers, still visible, with no erosion between layers defying claims of creation millions of years ago. Rather, this points to the actions of flowing water or hot flowing lava.

Carbon Deposits, massive amounts everywhere the evidence of instant burial of animal and vegetable matter. Pressure from carbon gasses still not fully released impossible to contain for more than a few thousand years.

Cambrian Explosion, named this way, but evidence of the instant demise, burial grounds for a wide variety of fossils all around the world.

Leap Seconds, rotation of planet Earth slowing yearly previously described. This provides evidence of the lifecycle of our planet in thousands of years only, not millions or billions.

Relationships, are complementary actions between male and female creating new offspring for humans, and all living creatures. There is no evidence of anything evolving now or in the past.

Inter-Dependence, multiple examples of very necessary dependence, essentials for a living planet; carbon dioxide from humans and all living creatures for plant growth; oxygen produced by plants for all living creatures. This suggests that both must be in existence at the same time, no eons of time possible for each to evolve separately as the "Evolution Theories" demand.

Therefore, I leave these many questions for you to ponder. May I suggest that all you require to verify this information or come to a logical conclusion is to access the reasoning power available to each of us!

References:-

1. Uniform Action-Uniformitarianism This page was last modified on 3 August 2014 at 13:46. 04August2014/ <http://en.wikipedia.org/wiki/Uniformitarianism>

2. Principles of Geology This page was last modified on 18 February 2014 at 04:34. 04August2014/ <http://en.wikipedia.org/wiki/Principles_of_Geology>

3. Ibid.

4. Geological Time Scale This page was last modified on 27 July 2014 at 16:47, 01August2014/ <http://en.wikipedia.org/wiki/Geologic_time_scale>

5. Eocene epoch World of Earth Science. 2003. Encyclopedia.com. 05September2014/ <http://www.encyclopedia.com/ topic/Eocene_epoch.aspx>

6. Theistic Evolution This page was last modified on 27 July 2014 at 01:58. 01August2014/ <http://en.wikipedia.org/wiki/Theistic_evolution>

7. Transitional Fossils -Missing Links This page was last modified on 29 July 2014 at 19:49. 01August2014/. <http://en.wikipedia.org/wiki/Transitional_fossil>

8. Search for Extraterrestrial Intelligence-SETI This page was last modified on 24 July 2014 at 08:18,01Aug2014/ <http://en.wikipedia.org/wiki/SETI>

9. Herbert Spencer This page was last modified on 22 July 2014 at 19: 01August2014/ <http://en.wikipedia.org/wiki/Herbert_Spencer>

10. Big Bang This page was last modified on 24 July 2014 at 00:04, 01August2014/ <http://en.wikipedia.org/wiki/Big_Bang>

11. River Deltas This page was last modified on 4 August 2014 at 01:39.01August2014/ <http://en.wikipedia.org/wiki/River_delta>

Chapter Seven

Micro to Macro Theory

Look at you! You are magnificently made, not even just made, mysteriously born, all started with two cells, a relationship between male and female, connected by an act of love, producing a new creation. This includes a concerted plan to extend love to the new offspring.

You are not alone, there are nearly seven billion (7,000,000,000) of us on this planet, as of this writing, the only planet designed to support every minute need of a living creature. Living needs for now, or may be required in the future.

Amongst the seven billion others like us, there are small "Micro" differences in colour of skin, height, weight, and features, but we all have the same blood within fixed blood types. This gives us an ability to exchange blood with each other, a distinct human blood. We are of the "human kind" distinct from any other "kind" of creature on earth.

We reproduce very fast, in the last one thousand years multiplying from a population of 310 million to one of 7 billion[1], in spite of war or disease, giving further evidence that our unique existence goes back only a few thousand years. The nearly seven billion humans now living, plus multiple billions before us you can be certain, were born from two cells. One cell from male, the other from female, the method of our identical beginning, each one with unique DNA[2], with unique finger prints. Creation is often attempted, a combination of chemicals in a laboratory without a single successful case to report.

We now know that all humans had to arrive here from an original male. The male sex contribution includes a unique Y-Chromosome[3], passing this down to the next male, never to a female birth. This fact confirms that no human offspring is possible without an adult, male and female. In fact, this applies to every living creature on the planet. The unique DNA of one kind of male cannot merge with that of a different kind of female, there must always be a "match" to reproduce offspring.

Southern Asia is the logical location of that first male, a continent with the largest population, known as the "cradle of civilization[4]." That we migrated from Africa, "millions of years ago," as some have proposed, is not readily apparent. The features of the African race are so different from the Asian race. In addition, the number of our present population does not support such a proposal.

There is no evidence that we evolved from a different kind of species, or even plant form, the claim of "Macro" evolution. The "missing links[5]," otherwise known as transitional forms, still missing from the fossil record in spite of extensive search since this idea was proposed. Most fossils are sea creatures providing evidence of sudden death commonly called the Cambrian Explosion Period[6]. Rather than an indication of sudden death; evolution thinking assumed a massive explosion of life 350 million years ago.

Many of these fossils are identical to present day live ones, no "Macro" evolution change during that long period! Does the evidence give us a very different picture?

The difference between micro and macro evolution is a major point of confusion between the original creation worldview and the Darwinian "Deep Time[7]" evolutionary worldview. Microevolution is the adaptations and changes within a species sometimes called mutations[8]. Macroevolution is the addition of new traits or a transition to a new species as proposed by the "Evolution Theories." Microevolution is a fact, observable throughout nature, always within that kind of species. Macro evolution is a theory without evidence in nature in spite of extensive research. If there is a point of contention, it is usually what one chooses to believe rather than what is evident.

Micro evolution is a fact as previously stated. There is no dispute for this fact. It is alteration of a specific trait due to natural response. Darwin's observation of the changes in finches, isolated in the Galapagos Island, show finches there that had much longer beaks than those found off the island. His assumption was that "Deep Time[9]" was changing a species into a different one. However, these finches remained finches! Darwin would observe this without sailing to the Galapagos Islands, an extensive variety of species such as dogs, cats, cattle, horses, etc. These variations are a result of breeding by humans but these variations can also occur as nature adapts to environmental change. Humans are no exception with a wide variety of features but still "human." The many suggestions of evolution are that species will change slightly over time to evolve into a different kind becoming a new species. This theory first proposed as a hypothesis then adopted. No scientific facts found to support it, no evidence in nature is evident, and the fossil records have zero transitional forms. Fossilized insects such as spiders and ants found, dated to pre-historic times, are identical to modern day spiders and ants.

The idea of a half-evolved creature seems impossible. For example our bodies need all their parts operating simultaneously. This need for "completeness" clearly observed from the most primitive single celled animal to the most complex mammal. This clearly contradicts Darwin's principle of natural selection. Even with mutation, skin will still be skin, and eyes will still be eyes. Different genes can create distinct variations but there is a limit. There can be rapid changes but inevitably there is a return to the norm. "The upper and lower limits to which mutation rates can evolve are the subject of ongoing investigation[10]."

This chapter started with a human "look at you," we know that we cannot live without any one organ or part. No partial human has ever been born amongst the human race or found in the fossil record with seven billion living and multiple billions returning to their original state, namely dust!

This subject can involve a very technical explanation in support of Micro only, relatively small changes within a "kind" of species. The fact remains that there is no evidence for Micro to Macro progression. Evidence supports progress rather than the random chance of Evolutionary Theories. We see various forms of progress such as the automobile, airplane, computer, telephone, etcetera, progressing over time, never by chance, and a result of human creative ability. This progress carried out by intelligent creative humanity, executing a well laid out plan.

Therefore, I conclude that the human race is very different from any other living creature, given an ability to create. We are so different from any other creature, not evolving from a simple atom or inert chemicals!

References:-

1. World Population This page was last modified on 31 July 2014 at 20:33. 02August2014/ <http://en.wikipedia.org/wiki/World_population>

2. DNA This page was last modified on 16 July 2014 at 18:52.02August2014/ <http://en.wikipedia.org/wiki/Dna>

3. Y- Chromosome This page was last modified on 2 August 2014 at 13:47. 02August2014/ <http://en.wikipedia.org/wiki/Y_chromosome>

4. Cradle of Civilization This page was last modified on 19 August 2014 at 13:12. 05September2014/ <http://en.wikipedia.org/wiki/Cradle_of_civilization>

5. Transitional Forms - Missing LinksThis page was last modified on 29 July 2014 at 19:49. 02August2014/ <http://en.wikipedia.org/wiki/Transitional_fossil>

6. Cambrian Explosion This page was last modified on 20 July 2014 at 02:56.02August2014/ <http://en.wikipedia.org/wiki/Cambrian_explosion>

7. Deep TimeThis page was last modified on 7 June 2014 at 23:27.02August2014/ <http://en.wikipedia.org/ wiki/Deep_time>

8. Mutations This page was last modified on 28 August 2014 at 18:50. 05September2014/ <http://en.wikipedia.org/ wiki/Mutation>

9. Ibid.

10. Mutation Rate This page was last modified on 30 July 2014 at 09:17.02August2014/ http://en.wikipedia.org/ wiki/Mutation_rate

Chapter Eight

Supernatural Recorded Events or Natural Selection Theory

The actions of birth are supernatural growing from two cells to trillions of cells[1] in nine months. Each has detailed instructions, instructions on how to build another human. The process of birth repeated billions of times for humans, multiple trillions for other creatures, all part of creation. Humans born according to their kind, humans producing humans, dogs producing dogs, cats producing cats, etcetera, all according to their kind. History has never recorded any "bird" born of a human, or reptile, or dog, or cat, or any other kind other than another human. Humans are born in nine months, no evidence that it takes billions of years, an action of love between a man and woman, love passing to the next generation! Love for each of my children when they were born, remaining with me, a parent.

I was fortunate to see my last two children born, the presence of a father not encouraged before in the delivery room. I witnessed a supernatural event; seeing at least one small part of this miracle. A baby is born, one minute living in a sack of fluid, the next minute breathing air with minimal intervention on the part of the doctor. How can one understand with our finite mind, every intricate organ perfectly made, a complete body with pumping heart and veins carrying the blood of life to every part of our body, to mention just a few of the many supernatural events. Every detail included in the plan, which we now call part of the contents of DNA[2] with many instructions. I further explain what I witnessed!

It is only a minute detail, but one most necessary for that embryo to exist in the mother's womb. This may be a detail to me but certainly requires prior knowledge and planning.

"What is all that white material covering the baby's body when born?" I enquired from the doctor.

"That is called vernix caseosa[3]," the doctor then explained. "It is a special material designed to protect the delicate skin of the baby when in a sack of fluid. White material keeps the skin from destruction in a fluid for nine months essential for life of a baby and vital for human survival!"

I quickly mention just one more, the indescribable love that welled up inside me, the love of parent for child, for nurturing and protection. This too is hard to explain, not being physical in nature, but rather spiritual. If one has doubts concerning supernatural occurrences, surely that notion will soon disappear while witnessing the miraculous birth of a new creation and the feelings associated with it!

Now we come to a much larger birth, our planet Earth with supporting solar system. Is this a most unusual occurrence, an integrated design which continues to exist, or just a chance happening as the following tale suggests?

Once upon a time, billions of years ago, some inert* chemicals came together, decided to create the first human *complete with all living creatures. They also threw in some plants to feed them. However, before that, they "banged out" a planet or two along with the solar system, a dwelling place for all life forms that evolved, complete with a universe. Therefore, they did it and they all lived happily ever after!*

*The Wikipedia Encyclopaedia definition of inert described as follows; "In English, to be inert is to be in a state of doing little or nothing."

This is obviously a fairy tale version of the original evolutionary story conceived in the 18th and 19th century. The idea of an explosion, a natural event, happening billions of years ago, creating all, is a more recent suggestion. It does seem like a fairy tale version, one of many trying to explain a "beginning." What started all of this is a constant question for science coming up with an array of theories.

A comment from a notable scientist however seems like a logical conclusion. I first heard this suggestion from a movie produced by Ben Stein called "Expelled – No Intelligence Required[4]." The movie starts with several interviews of scientists who chose to disagree with the "Evolution Theories." The movie shows different methods used to censor their opposition to the "status quo." Toward the end of the movie, Ben Stein interviews Richard Dawkins a confirmed advocate of the Darwinian Theory of evolution. In the interview, this scientist states he did not believe in any God, that idea is a primitive superstition invented by man. When Ben Stein asked about creation of heaven and earth the reply was that it was a very slow process, started by a self-replicating molecule. A civilization evolved by some form of Darwinian means with a very high level of technology, designed a form of life that they seeded on this planet.

Ben Stein concluded that he was talking about an "intelligent designer" from a different planet, a natural occurrence, not the Creator of Genesis, from Jewish history.

Men of great influenced in the late 18th and early 19th century convinced the scientific community to ignore any supernatural even as described by Moses in the Jewish Torah, the Old Testament. The influence of just a few men dominates the scientific community even today in spite of more recent knowledge. In most schools around the world this is the only acceptable method of scientific research, any supernatural events considered religion or mythology. The theoretical idea coming from the scientist previously mentioned was in line with a decision made ages ago to ignore the supernatural. In a subsequent statement, Ben Stein suggested that we might experience God from science if we have the freedom to go there!

In principle, I can accept the idea that Richard Dawkins gave as an explanation of the beginning of earth. This suggests that a civilization with a higher level of technology designed the many forms of life found on our planet. Some choose to call it an Alien meaning someone foreign to us, someone from a different planet or place. The story from this scientist indeed includes an Alien from a more advanced planet. That Alien dropped a seed or seeds here several billions of years ago, then before that, another Alien from a more advanced planet dropped a seed on that planet, and so on, ad infinitum. This development of other more advanced planets in the same way described by the Darwinian Evolutionary Theory. More advanced Aliens on other planets is the suggestion. I believe that Richard Dawkins was right because recent scientific knowledge involving DNA[5] indicates that any seed or cell is much too complex to evolve by itself. It must come from some other planet or place, an intelligent design by a more advanced civilization. The part I have trouble believing is that any intelligent entity would do it this way, over a period of billions of years, eons of time.

This is my reasoning! Our planet provides evidence that all life forms start from adults only in a very precise manner, involving a female egg with male sperm. The resultant embryo grows in a protected environment following a very complex set of instructions in the human seeds placed there from some intelligent source. An outside entity is necessary with "prior knowledge," knowing the precise needs of the new species, in this case a human. To drop any material for future development on another, possibly very hostile planet seems like a strange suggestion. To consider that it would take billions of years to develop those trillions of cells complete with billions of instructions seems very remote. Would an intelligent being do it this way? When Europeans discovered the New World, adults moving there then produced offspring from adults who started a colony there.

If some outside entity found this planet, would it leave seeds or adults to form a new colony similar to the one familiar to them? Why leave "seeds" that would take billions of years to evolve under questionable conditions?

If intelligent beings came here, the planet would have to meet all the conditions necessary to support their particular needs. If it were a hostile environment it seems logical that they would use their superior intelligence and ability to make the necessary alterations. Experience indicates that those changes would need to follow a very logical sequence of events.

In building a home, one follows an architectural plan which includes a sequence of construction; the roof not built before the foundation. If a visitor from another planet prepared our planet for complex living creatures, a very precise sequence of events is necessary, location the first consideration. If we examine the Genesis story, we find a very logical sequence of construction events!

A planet is prepared for a variety of life forms complete with the addition of plants to feed them. It states that an entity from a different place, with a Spiritual body able to hover, prepared our watery planet using knowledge and ability.

Do we continue to dismiss this account because men of great influence over 150 years ago told us to? Were we convinced to ignore any supernatural event in our discovery of the truth?

Before I describe this logical series of events, preparing our planet for life, I would like to describe a serious ongoing effort to discount any supernatural events. Some say the account of a "beginning" in the book of Genesis is not true history but rather mysticism, mystery, or myth told to the Jewish people, years ago, to satisfy their curiosity. Others say that it is not Hebrew narrative prose but rather just Hebrew poetry[6]. Still others suggest that it requires interpretation by a highly educated individual. Is all this an attempt to support the idea that all evolved because of natural causes instead of a creation event?

The first sentence of the Genesis story confused me when it says "heavens," plural rather than "heaven" singular. What do we mean by the word "heavens?" Are there several heavens?

Finally, I decided to consult an original Jewish bible written according to the Masoretic text[7]. There the word is "heaven" singular from Genesis 1:1 later described as "firmament" in

Genesis 1:8, the area just above the earth, the area containing water. This water later described as descending during a Global flood. The same singular use of "heaven" found in Exodus 20:11, part of the Ten Commandments. These events in Genesis and Exodus history discounted by science as supernatural, changing a previously accepted world view. I found where the change from singular to plural of "heaven" may have originated by consulting the Wikipedia Encyclopaedia in the following text under "Firmament[8]," indicating a decision to change the original Jewish text including the reason:

"Conservatives and fundamentalists tend to favor translations that allow scripture to be harmonized with scientific knowledge, for example "expanse." This translation used by the New International Version and by the English Standard Version. The New Revised Standard Version uses "dome," as in the Celestial dome."

New versions of an old original account changed to "heavens?" My inquiry continued remembering what the Apostle Paul wrote in 2 Corinthians Chapter 12: Verses 1 to 4. Verse four of the New Testament Bible describes a "third" heaven as Paradise. My curiosity was aroused, a logical question, where is the second heaven? I had found the first heaven defined in that Jewish Bible as the "firmament" around the earth. To get an answer I turned to the same source of information coming from Paul. The Apostle Paul's writings have credibility much more than a small group of influential men when it comes to historical accuracy. This is why! For those unfamiliar with this famous teacher allow me to explain.

Paul is the author of a major part of the New Testament. He was born in Tarsus an important Roman town in Asia Minor, educated in Jerusalem, "at the feet of Gamaliel, according to the perfect manner of the law of the fathers," as stated in the Book of Acts. Gamaliel is a descendent of a long line of Rabbis, teachers, called Pharisees, studying the scriptures for centuries, including the writings of Moses. Paul; a student of Gamaliel is considered an exceptional student.

Initially Paul persecuted the followers who had witnessed or believed in a man called Jesus as their Messiah. Paul even had many followers killed, determined to eradicate a sect whose beliefs seemed to conflict with Jewish tradition. However, a supernatural event changed this man, realizing that Jesus fulfilled all the requirements of a Jewish Messiah, foretold by their Prophets centuries ago. The Prophet Isaiah even gave a detailed account of the method of death for Messiah in Isaiah Chapter 53. Paul had serious credentials living much closer to original events learning from teachers going back in time. Learning about the writings of Moses, someone educated in the Royal courts of Egypt. Writings started in Mesopotamia only a few decades before, a nation called the oldest civilization, in a Fertile Crescent. This area includes the Tigris and Euphrates Rivers at the foot of mountains including Mount Ararat where Noah's Ark landed. I believe that Paul had sufficient credentials to state that there is a "third" heaven. Second heaven now understood by reading the account of Genesis with a renewed understanding. The second heaven is an area between the first heaven, "Firmament", and the third called "Paradise," as referred to by the apostle Paul.

In addition, I would not be satisfied with my analysis of Genesis unless I knew who wrote it. Was it just fable or Jewish poetry as some suggest? In Jewish circles, the first five books of scripture, Moses writes the Torah. Those writings include many accounts of supernatural events, certainly not accepted by modern science. Again, I decided to consult Wikipedia Encyclopaedia under the word "Torah[9]":

"According to religious tradition, all of the teachings found in the Torah, both written and oral, were given by God to Moses, some of them at Mount Sinai and others at the Tabernacle, and all the teachings were written down by Moses, which resulted in the Torah we have today. According to a Midrash, the Torah was created prior to the creation of the world, and was used as the blueprint for Creation."

Then it goes on to state:

"The majority of Biblical scholars believe that the written books were a product of the Babylonian exilic period (c. 600BCE) and that it was completed by the Persian period (c. 400BCE)."

There continues to be an attempt to find an understanding of a beginning by natural means only in spite of contrary evidence. No supernatural event is ever considered! One of the latest theories includes a "big bang" explosion at the beginning. To suggest that an initial explosion created all things is indeed a mystery in itself that requires further explanation of how it got started and who did it! As our knowledge progresses is it possible to suggest that we simply evolved from some prehistoric chemical soup. Knowledge of the complexity of a single cell, called DNA[10] is now common knowledge. In spite of this there continues to be an attempt to explain our beginning in terms of this Natural World. The result seems to be only imaginary suggestions or theories.

The creation of all that we are able to see and experience calls for some Spiritual form of existence. This Spirit needs to come from an existence foreign to us. The intricate requirements are beyond the ability of nature. A closer look also reveals extensive symbiotic[11] relationships, an inter-dependent need at the beginning. This suggests the need for immediate action rather than gradual change with no interdependent connection over time.

The Genesis story alone fits this need. It starts with a brief description of the presence of a Spirit form, unlike us, with an ability to create. Genesis, story of the beginning, starts with a Spirit hovering over a water-filled planet. I use the term Spirit which means someone from a foreign location, in a form much different from us, our body only designed for this planet. This entity had a Spiritual body able to hover. This Spirit decides to transform the watery object into a planet that supports life forms, including a wide variety of plants to feed them. Details or methods of construction are not included but the events follow a very logical sequence. I purposely chose a translation of Genesis that is close to the original document as found in a Jewish translation. I include a small sample of that Jewish document at the end of this chapter describing creation, written in a manner that all generations can understand. The descriptive sequence of events follows, also using our recent knowledge:

Location: the original Earth chosen by the Spirit was unformed, void but with lots of water. The word Heaven, singular, mentioned in the first sentence later described as firmament. The planet consisted of earth and water with a firmament surrounding it. Darkness is everywhere!

Preparation: the first step is to bring light to the darkness provided by the Spirit. We now learn that this light is directional, necessary for day and night; there is evening with morning on a rotating planet. A planet with an abundance of water, with some of the water used to create a canopy in the upper part of the firmament, possibly a very thick layer of water vapour. A canopy above the planet is protection for life forms from adverse sunrays, included in creation events that follow. Planet earth continues to revolve on the first and second days.

Provision: the earth has plants, in addition to sea creatures, which come next. A variety called "kinds" of seeds planted in the fertile earth to feed humanity including a variety of land and sea creatures. These are events of the third day.

Maintenance; the sun, moon, and stars are introduced on the fourth day. This forms part of the Second Firmament. Special Sunrays, rare among stars providing photosynthesis for plants to grow from seeds previously planted in the fertile soil. The moon churns the waters twice a day introducing air, oxygen, essential for sea creatures and plants. The moon provides stability for the earth's rotation while causing the planet to tilt 23°, an angle from the sun producing seasons essential for plants.

Life Forms: in the seas in abundance, land animals and birds flying in the first heaven below the canopy of water. All created to multiply sustaining a creation for all. A variety of life forms most necessary for the continuance of a living planet, all produced on the fifth day.

Inheritance: all is now ready for humanity to exist, man and woman created from the dust of the earth, "elements" found in our bodies. The Spirit breathed life in us. The first two humans, created maturely, a very necessary condition required for procreation, to fill the earth as commanded by the Spirit. Humans appeared on the sixth day, then given an inheritance, dominion over all creation.

Rest: is an example for humanity to rest from work every seven days. This is a rest period to take time to thank the Creator Spirit for all done for us, a time to learn how that Spirit wants us to live. We have an ability to choose, use our creative ability with freedom.

However, we chose to believe a lie, causing the first death, inflicting pain and sorrow on future generations, including all creation. A creation called "very good[12]" The Spirit did not just create but gave us a set of rules of how we are supposed to live for our protection. Ignorance of these rules in the 20th century resulted in suffering and death of millions of lives, massive destruction and wasted resources. This was the result of unproven ideas and the opinion of only a few people, abusing their positions of influence.

The Jewish historical record describes a nation, Israel, a people who worshiped the Creator but then chose to worship idols made of wood and stone, ignoring warnings from their Prophets. This was the practice of nations around them, which they chose to copy. This led to disastrous results all carefully explained in Jewish history, called the Old Testament. History repeats, today we choose to believe that all came from the natural world, nature, rather than consider any Creator or creative process. This is similar to what the nation of Israel did in the past. We have not learned from history!

It is interesting that the Founding Fathers of a new nation, the United States understood the valuable lessons from Jewish history. Their constitution includes reference to a Creative Spirit. This recognition is included in their many quotes. One of these quotes comes from Daniel Webster[13]:

"If religious books are not widely circulated among the masses in this country, I do not know what is going to become of us as a nation. If truth be not diffused, then error will be. If God and His Word are not known and received, the devil and his works will gain the ascendency. If the evangelical volume does not reach every hamlet, the pages of a corrupt and licentious literature will. If the power of the gospel is not felt throughout the length and breadth of this land, anarchy and misrule, degradation and misery, corruption and darkness will reign without mitigation or end."

Therefore, did we experience some of the disastrous results mentioned by Daniel Webster in the recent past? A following chapter provides an example of what can happen when we chose to ignore the supernatural, thinking we are all powerful in our "finite" existence and knowledge to change history.

Which story has more supporting evidence, the history behind "Creation" or the many "Evolution Theories?" I leave these questions for you to consider!

Partial Masoretic Jewish Biblical reference:

Genesis 1:1 "In the beginning God created the heaven and the earth."

Genesis 1:8 "And God called the firmament Heaven. And there was evening and there was morning, a second day."

Genesis 1:9 "And God said: 'Let the waters under the heaven be gathered together unto one place, and let the dry land appear.' And it was so."

Exodus 20;11 "for in six days the Lord made heaven and earth, the sea and all that in them is, and rested on the seventh day; wherefore the Lord blessed the Sabbath day, and hallowed it."

References:-

(All Biblical references are from the New International Version unless otherwise stated.)

1. Cell This page was last modified on 26 July 2014 at 03:49. 02August2014/ <http://en.wikipedia.org/wiki/Cell_(biology)>

2. DNA This page was last modified on 16 July 2014 at 18:52.02August2014/ <http://en.wikipedia.org/wiki/Dna>

3. Vernix Caseosa This page was last modified on 27 July 2014 at 01:25. 02August2014/ <http://en.wikipedia.org/wiki/Vernix_caseosa>

4. Expelled-No Intelligence Required This page was last modified on 14 July 2014 at 11:10. 02August2014/ http://en.wikipedia.org/wiki/Expelled:_No_Intelligence_Allowed>

5. Ibid

6. Hebrew Poetry This page was last modified on 16 June 2014 at 21:07. 02August2014/ <http://en.wikipedia.org/wiki/Biblical_poetry>

7. Masoretic Text This page was last modified on 4 March 2013, at 16:55.02August2014/ <http://www.newworldencyclopedia.org/entry/Masoretic_Text>

8. Firmament This page was last modified on 7 July 2014 at 14:48.02August2014/ <http://en.wikipedia.org/wiki/Firmament>

9. Torah This page was last modified on 1 August 2014 at 05:56. 02August2014/ <http://en.wikipedia.org/wiki/Torah>

10. Ibid.

11. Symbiosis This page was last modified on 3 September 2014 at 05:47. 06September2014/ <http://en.wikipedia.org/wiki/Symbiosis>

12. Very Good, Genesis 1:31 New International Version "God saw all that he had made, and it was very good. And there was evening, and there was morning--the sixth day."

13. Daniel Webster © 2014 Goodreads Inc., 02August2014/ http://www.goodreads.com/author/quotes/31180.Daniel_Webster

Chapter Nine

Denial of Intelligent Design

Can I open it and look inside? I was afraid father would say "no" then I would die of curiosity, probably get into trouble for taking it apart, looking inside anyway. That curiosity had gotten me into trouble many times before.

"Okay Stanley," Dad finally replied. "If you think you can take it apart, and don't leave a mess around for your mother to pick up, and then put everything back in the shed where you found it."

It was round and shiny, sitting on two impressive looking little legs, with a lever like leg behind that kept it up right. It had a strange ticking sound, at night the numbers lit up in the dark along with the two hands that swept round. I know its main purpose was to wake mother and father in the early hours of the morning with its loud ringing, but now it was silent.

That shiny silver cover was difficult to remove but my stubborn persistence paid off! The insides were golden with all sorts of little gears and levers that were most impressive! Why did it refuse to tell time anymore, eventually to produce that loud ringing noise? I stared and stared at this confusing mess of strange workings.

"This is much too difficult for me to ever understand," I thought, "but also much too fascinating for me to let go." I would stop to stare at it many times then go away with an irresistible urge to look inside once more. Each new glance seemed to offer another clue.

Then, one day, I saw it! A strange little piece that had lost its golden colour, the ends had silvered with the wear of it.

"What could this mean?" I went away thinking, trying to solve the mystery. Who am I to think I could fix such a device at my age when father was too busy to find the time to do it himself. I knew Dad could fix it though, he was my father, and he could fix anything!

Then one day, I took it out once more, examining that worn silver thing. Worn and bent by doing too many "tick tocks." I bent it back and forth several times until finally to my great surprise the old clock started to "tick tock" once more. "I did it!"

The shiny silver cover was very difficult to replace with screws forever rolling onto the floor. Finally, all pieces clicked into place and there stood the shiny clock once more on those impressive little legs. The two "glow in the dark" hands were all askew! This condition was quickly fixed by checking with the big clock in the kitchen. The "winding device" behind was then turned until tight.

"Stanley, I thought you were going to take that clock apart? Why is it still in one piece?" My father had finally arrived home.

"I did, I did, take it apart," I was quick to explain.

"It works again, fixed, see!"

Father picked up the old clock looking at it carefully, shook it, listened to it, and then put it down on the table where he found it. I do not remember him saying anything, but never forgot the

85

look on his face as he stared at me in amazement for what seemed a very long time.

This clock, including many parts, complete with spring and pendulum, is evidence for design. It did not evolve from a sundial!

When our family moved to a farm, my curiosity was most active. That cream separator bothered me. The turning of the handle, just right, caused a distinctive ring, me turning it at every chance. Mother told me to stop but the contraption was too fascinating. Raw milk, still fresh from the cows poured in the top, reduced milk out one spout, cream from the other.

Yes, I took it apart, when no one was looking! The astonishing insides of a machine, with its rapidly spinning parts, powered by a turning handle, an example of centrifugal force, lighter milk separated from heavier cream. A planned design of a special machine based on "prior knowledge" from someone coming up with this idea!

Then I could not leave that old Model T. Ford truck alone for long. Father finally explained a tragic event, the demise of a very prosperous wheat farm in Manitoba. The depression took its toll reducing our family to poverty.

"We all traveled east in that old truck, your mother and I, with your three older brothers and Grace just a baby then," Dad explained.

Oh, so many experiences on the farm, tractors, thrashing machines, mowers to name just a few, all carefully examined. I saw what Ford, Massey, Ferguson and others were intelligent enough to invent. That farm helped satisfy my curiosity, teaching me much!

The animals on the farm kept me busy. I watched calves born, sometimes in the early hours of morning and oh, what I learned from those chickens. A question has perplexed humanity for centuries, about chickens!

"What came first the chicken or the egg[1]?"

That question can soon lead us to "circular reasoning[2]." I found answers on that farm. The rooster and hen had to come first which involves a "built in" irresistible attraction between rooster and hen! There are no chickens or eggs without a relationship first. Rather, I suggest a different question, "What came before the rooster and hen and what created that built in attraction?"

However, I soon learned; do not mess with a rooster! If you annoy his "hen friends" or try to steal their eggs under them you may feel Mister Rooster landing on your back, fangs extended, furiously pecking at your head! An egg becomes a chick only with a contribution from that nasty rooster first!

The farm taught me that all creatures born whether human or chicken involves a relationship between male and female. There is also a built in instinct to protect their offspring. That farm provided no evidence of anything ever evolving by chance!

As far as "circular reasoning[3]" previously mentioned concerning that famous chicken and egg question, apparently it is only an acceptable form of reasoning when one ages rocks and fossils. Certain "Index Fossils[4]" used to determine the age of rocks, then "Type of Rock[5]" used to age the "Index Fossils[6]" as indicated in the following quotation:

"The intelligent layman has long suspected circular reasoning[7] in the use of rocks to date fossils and fossils to date rocks. The Geologist has never bothered to think of a good reply."[8]

There were excellent educators on that farm as well, much education for me.

Two sisters a source of much learning for their younger brother, while he, for them, a most valuable source of entertainment. I received a massive dose of that education on one particular Sunday afternoon, something I still vividly remember. These two wonderful educators told me that our father wanted that frisky calf taken into the barn. "We know that you can do it Stanley," they reassured me.

"Oh they have so much confidence in me," I reasoned, my pride pumped. I was ready for the challenge to the delight of those two mischievous sisters. I quickly found a rope finally securing it around that calf located in the corral.

It took less than a minute before the calf had a different plan. Once roped, my victim proceeded to head for the open pasture to be with mother. Well, my lesson started then as I soon learned that cows do more than just graze on a lot of grass in pastures. That frisky calf very soon had me in a horizontal position dragging me through very unpleasant material. That rather smelly material covered me from head to toe. This stubborn little boy refused to let go, again, to the delight of those older sisters. They had their amusement on this otherwise very quiet Sunday afternoon. I admit that these sisters were not alone with their mischief; I certainly provided enough of my own. Cousin John can vouch for that!

That incident taught me something about "common sense." Sometimes it is better to let go of your pride or you may find yourself over your head in rather unpleasant material. In addition, do not believe everything that you hear, but rather check the source with an eye to a possible reason behind the deception as I should have done with my sister's story!

Experiences in life led me to much interest in science, providing fascination for me with the Evolution Theories. Knowledge has accelerated mainly in the 20th century since those theoretical ideas first arose placing many doubts in my mind, later learning that I certainly am not alone.

One brave enough to express concern is Michael Behe[9] a biochemist, professor at Lehigh University in Pennsylvania. Using an electron microscope not available when the Evolution Theories were born, he made a most important discovery. Examining flagella bacteria[10] with the electron microscope, he concluded that this type of bacteria could not exist according to evolutionary rules of gradual change over time. All parts are necessary at the same time for its existence! A discovery he called "Irreducible Complexity[11]" outlined in his book, "Darwin's Black Box[12]."

Michael Behe[13] is willing to question Darwinian principles! He proved that small changes from nature's accidental design over a period of "Deep Time[14]" were not a logical conclusion. These statements caused much alarm from the established scientific and academic community indicating displeasure with this scientist's suggestion. It became obvious that his discovery did not fit the accepted scientific theories.

This scientist's radical new idea suggested "Intelligent Design[15]" rather than "Natural Selection[16]" over eons of time. This is not a new idea, but Behe[17] was brave enough to state it. Darwin himself when he considered the complexity of the eye realized that it was "irreducibly complex[18]" as well. This is what Darwin stated about the complexity of the eye in his book "On the Origin of Species by Means of Natural Selection, or the Preservation of Favoured Races in the Struggle of Life," 1859, Chapter VI,

"To suppose that the eye, with all its inimitable contrivances for adjusting the focus to different distances, for admitting different amounts of light, and for the correction of spherical and chromatic aberration, could have formed by natural selection, seems, I freely confess, absurd in the highest degree possible."[19] Charles Darwin

In the same way, can we assume that there is no "intelligent design[20]" with our solar system, no designer? Is our solar system the result of "Natural Selection[21]" or some "Big Bang[22]" happening billions of years ago? I submit that just a few facts soon dismiss such an assumption, that nature alone or a massive explosion ever created such an engineering marvel.

This is my reasoning! Our knowledge has progressed to such a degree that we can now launch a variety of satellites from powerful rockets rising from our planet. Satellites weighing up to a few tons circle the earth, actually falling around it, conforming to the established laws of gravity. The rocket must reach a certain speed to accomplish this amazing objective. Too fast, the satellite continues, entering space, too slow, it plunges back to earth. Our planet earth is also a satellite 38,624 kilometres in diameter (24,000 miles), rotating 1,609 km/h (1,000 mph), traveling 156,106 km/h (97,000 mph) in a journey around the sun once a year.

What kind of rocket is required to launch such a massive satellite? What launched it with such precision and engineering perfection?

This massive object Earth falls around the sun in an elliptical orbit, too close to the sun, we fry, too far, we freeze. An elliptical orbit designed to circle the sun to support life on our planet. In addition, the moon demonstrates a nearly perfect circle around the Earth also essential in support of life. Two objects we accept as necessary, but how many others are still undiscovered?

Then there is prior knowledge, architects design of a home is one example. The architect naturally starts with an intricate knowledge of the needs of all the inhabitants. Can a production as intricate as our solar system just happen by chance without any prior knowledge yet provide for all the needs of a multitude of life forms? Just the moon alone is an absolute necessity, an ideal size, exact distance, proving an essential interaction with earth. The gravitational effect keeps our planet from wobbling while producing an exact 23-degree angle from earth's axis, essential for seasons.[23]

Another essential action from the moon is similar to the needs of fish in a home fish tank. There must be air bubblers and plants included in that tank, keeping fish alive. The moon's gravity causes tides[24] twice a day acting as a massive bubbler for sea creatures living in oceans. This demands an exact size with gravity from the moon in relation to the size of planet earth to accomplish this most essential action.

In addition, consider the following concerning our planet within the solar system as described in a DVD called "The Privileged Planet[25]":

Location: in an outer edge of our galaxy, not central, away from hostile space debris. It includes several giant planets on outer edge of the solar system, a shield for our protection. Earth's magnetic field[26] protects us from dangerous rays coming from outer space. A unique clear window in an otherwise very crowded Milky Way Galaxy[27] encouraging us to explore the outer reaches of the cosmos.

Water/Food: is a provision essential for life forms. This is only possible according to an exact orbit sometimes called the "Goldilocks principle[28]," not too close, not too far, from the sun. It includes an abundance of water necessary to absorb heat from the sun. A rare star, our sun, emitting unique rays for photosynthesis, needed to produce food. Different "kinds" of seeds planted in arable soil of earth to produce food essential for such a variety of life forms.

Atmosphere: with an exact mixture of Oxygen, Nitrogen, and Carbon Dioxide gases in support of complex life forms. This transparent atmosphere, unlike other planets, allows scientific discovery of outer space.

There are many more examples like this, with recent knowledge adding more to the list! The solar system is indeed an engineering marvel, an "integrated" system all designed to support humanity, all life forms, including all forms of vegetation to feed them.

Are we living on a planet that is designed for us, we designed for the planet? If we do not acknowledge the engineer, is it a gross insult?

We find support for this marvellous design in the Genesis account declaring all creation as "very good[29]," also including a tragic story as well. To explain I look back at the clock story of my youth suggesting a similar problem. The old clock had a flaw, hard to detect, but it grew eventually stopping the whole works. No more ticking, it was dead. The Genesis story relates how the first two humans refused to obey the Creator's direct instructions creating a flaw to an otherwise perfect creation resulting in death, a fate all of us have inherited.

Knowledge has increased to a point where we now have evidence that our planet is indeed affected. The rotation of planet Earth is slowing by slightly less than one second per year. It will not stop rotating in our lifetime, but the effects are evident. We have applied several seconds called "leap seconds[30]" to keep those atomic clocks in line with the slowing rotation of our planet.

Michael Behe[31] is not alone amongst scientists with his beliefs, but some scientists still claim there is no "Intelligent Design[32]" All actions not a result of intelligence but rather, they insist, by nature alone via "Natural Selection[33]" a product of Darwinian Evolutionary thinking.

Richard Dawkins[34] a world-renowned scientist and author is actively promoting this thinking, with books and worldwide lectures. He is an Atheist with no belief in any God claiming an undisputed belief in evolution as a fact not a theory. He makes a statement in his book "The Greatest Show on Earth" Chapter 4 called "Silence and Slow Time" concerning anyone who refuses to believe in evolution!

"If the history deniers who doubt the fact of evolution are ignorant of biology, those who think the world began less than ten thousand years ago are worse than ignorant, they are deluded to the point of perversity. They are denying not only the facts of biology but those of physics, geology, cosmology, archaeology, history and chemistry as well."[35]

This is quite a statement! Let's see what other noted scientists in those seven fields of scientific discipline have to say about that opinion! There are many examples; I choose only one from each, for brevity:

Biology - David Green and Robert Goldberger…

"However, the macromolecule-to-cell transition is a jump of fantastic dimensions, which lies beyond the range of testable hypothesis. In this area, all is conjecture. The available facts do not provide a basis for postulating that cells arose on this planet … We simply wish to point out the fact that there is no scientific evidence."[36]

Physics - Wolfgang Smith, PhD …

"The fact is that in recent times there has been increasing dissent on the issue within academic and professional ranks, and that a growing number of respectable scientists are defecting from the evolutionist camp. It is interesting; moreover, that for the most part these 'experts' have abandoned Darwinism, not on the basis of religious belief or biblical persuasion, but on strictly scientific grounds ….."[37]

Geology - Stephen Jay Gould …

"The absence of fossil evidence for intermediary stages between major transitions in organic design, indeed our inability, even in our imagination, to construct functional intermediates in many cases, has been a persistent and nagging problem for gradualistic accounts of evolution."[38]

Astronomy-Cosmology - Sir Fred Hoyle …

"The chance that higher life forms might have emerged in this way is comparable with the chance that a tornado sweeping through a junk yard might assemble a Boeing 747 from the materials therein."[39]

Archaeology - John Adler with John Carey ...
"The more scientists have searched for the transitional forms that lie between species, the more they have been frustrated."[40]

Natural History - Prof. J Agassiz ...
"The theory of the transmutation of species is a scientific mistake, untrue in its facts, unscientific in its method, and mischievous in its tendency."[41]

Chemistry - Prof. R Goldschmidt PhD ...
"It is good to keep in mind ... that nobody has ever succeeded in producing even one new species by the accumulation of micro mutations. Darwin's theory of natural selection has never had any proof, yet it has been universally accepted."[42]

One of the scientists not listed, Dr. T. N. Tahmisian even said the following:

"Scientists who go about teaching that evolution is a fact of life are great con-men, and the story they are telling may be the greatest hoax ever. In explaining evolution, we do not have one iota of fact."[43]

In addition, there are many Internet Websites with a list of notable scientists who testify that all we witness cannot just be the selection of nature without intelligent design and designer. The reference section of this chapter indicates one of these websites with quotations called "Scientifically Unproven."[44]

Therefore, leading scientists listed above leave their comments. However, we have the ability to come to our own conclusions by simple reasoning.

References:-

1. Chicken or the Egg This page was last modified on 11 August 2014 at 03:59. 11August2014/ <http://en.wikipedia.org/wiki/Chicken_or_the_egg>

2. Circular Reasoning This page was last modified on 6 August 2014 at 23:24. 23August2014/ <http://en.wikipedia.org/wiki/Circular_reasoning>

3. Ibid.

4. Index Fossils This page was last modified on 19 June 2014 at 00:34. 11August2014/ <http://en.wikipedia.org/wiki/Index_fossil>

5. Type of Rock (List of Rock Types) This page was last modified on 23 June 2014 at 11:20. 11August2014/ <http://en.wikipedia.org/wiki/List_of_rock_types>

6. Ibid.

7. Ibid.

8. (J. O'Rourke in the American Journal of Science, January 1976, Volume 276, Page 51)

9. Michael Behe This page was last modified on 23 June 2014 at 12:05, 07August2014 / <http://en.wikipedia.org/wiki/Michael_Behe>

10. Flagella Bacteria (Flagellum) This page was last modified on 15 July 2014 at 18:41. 11August2014/ <http://en.wikipedia.org/wiki/Flagellum>

11. Irreducible Complexity This page was last modified on 9 June 2014 at 19:43, 07August2014/ <http://en.wikipedia.org/wiki/Irreducible_complexity>

12. Darwin's Black Box This page was last modified on 23 June 2014 at 12:01. 07August2014/ <http://en.wikipedia.org/wiki/Darwin%27s_Black_Box>

13. Ibid.

14. Deep Time This page was last modified on 7 June 2014 at 23:27 07August2014/ <http://en.wikipedia.org/wiki/Deep_time>

15. Intelligent Design This page was last modified on 7 August 2014 at 03:18. 07August2014/ <http://en.wikipedia.org/wiki/Intelligent_Design>

16. Natural Selection This page was last modified on 18 July 2014 at 13:33. 07August2014/ <http://en.wikipedia.org/wiki/Natural_selection>

17. Ibid.

18. Ibid.

19. "On the Origin of Species by Means of Natural Selection, or the Preservation of Favoured Races in the Struggle of Life," 1859, Chapter VI, Difficulties, Page 85

20. Ibid.

21. Ibid.

22. Big Bang This page was last modified on 24 July 2014 at 00:04 07August2014/ < http://en.wikipedia.org/ wiki/Big_Bang>

23. Moon This page was last modified on 28 July 2014 at 20:51. 07August2014/ < http://en.wikipedia.org/wiki/Moon>

24. Tides This page was last modified on 24 July 2014 at 22:55. 14August2014/ <http://en.wikipedia.org/wiki/Tide>

25. The Privileged Planet a DVD produced by Illustra Media by Jay Richards and Guillermo Gonzales, Regnery Publishing

26. Earth's Magnetic Field This page was last modified on 10 August 2014 at 09:50. 11August2014/ <http://en.wikipedia.org/wiki/Earth%27s_magnetic_field>

27. Milky Way Galaxy This page was last modified on 5 August 2014 at 22:08. 11August2014/ <http://en.wikipedia.org/wiki/Milky_Way>

28. Goldilocks Principle This page was last modified on 26 May 2014 at 10:45. 11August2014/ <http://en.wikipedia.org/ wiki/Goldilocks_Principle>

29. Very Good, Genesis 1:31 New International version "God saw all that he had made, and it was very good. And there was evening, and there was morning--the sixth day."

30. Leap Seconds This page was last modified on 29 July 2014 at 08:27. 07August2014/ <http://en.wikipedia.org/wiki/Leap_second>

31. Ibid.

32. Ibid.

33. Ibid.

34. Richard Dawkins This page was last modified on 11 August 2014 at 16:56. 11August2014/ <http://en.wikipedia.org/wiki/Richard_Dawkins>

35. "The Greatest Show on Earth" Chapter 4 called "Silence and Slow Time" page 85

36. David Green and Robert Goldberger (1967) Molecular Insights into the Living Process New York: Academic Press, pp. 406-407 (David Green is a biochemist at the University of Wisconsin. Robert Goldberger is a biochemist at the National Institute of Health)

37. Wolfgang Smith in his book "Teilhardism and the New Religion: A Thorough Analysis of the Teachings of Pierre Teilhard de Chardin", Tan Books & Pub. Inc: Rockford (USA), 1988 p: 1

38. Stephen Jay Gould (Professor of Geology and Palaeontology, Harvard University), "Is a new and general theory of evolution emerging?" Palaeontology, vol. 6(1), January 1980, p. 127

39. Sir Fred Hoyle (English astronomer, Professor of Astronomy at Cambridge University), as quoted in "Hoyle on Evolution." Nature, vol. 294, 12 Nov. 1981, p. 105

40. John Adler with John Carey: Is Man a Subtle Accident, Newsweek, Vol.96, No.18 (November 3, 1980, p.95)

41. Prof. J Agassiz, "Professor Agassiz on the Origin of Species," American Journal of Science 30 (June 1860):143-47, 149-50

42. Prof. R Goldschmidt PhD, DSc Prof. Zoology, University of Calif. In Material Basis of Evolution Yale Univ. Press

43. Dr. T. N. Tahmisian (Atomic Energy Commission, USA) in "The Fresno Bee," August 20, 1959. As quoted by N. J. Mitchell, Evolution and the Emperor's New Clothes, Roydon Publications, UK, 1983, title page.

44. Scientifically Unproven This page was last modified on 8 February 2011, at 12:09. 08August2014/ <http://www.wicwiki.org.uk/mediawiki/index.php/Scientifical ly_Unproven>

Chapter Ten

Cause and Effect –
Survival of the Fittest

"Whatever makes men good Christians makes them good citizens"
Daniel Webster[1]

Steam came rolling out of the nostrils of those big Belgian work horses as they approached quickly, logs hauled on a large sleigh behind them, bound for the local train station, then destined to be used for building in Montréal, only 30 miles away. There sat Uncle Carl Carden atop the massive set of logs, frost on his eyebrows, moustache and beard, reins firmly held by those huge hands, as he guided the horses down the country road.

Carl was a favourite uncle, loved children; spoke encouraging words and always willing to provide that ride on his back. During the week, he rose early, made his way to the one-room schoolhouse each morning, stuffed a generous helping of wood to the pot bellied stove in the middle of the room, always making sure we entered a cozy warm room when classes commenced early the same morning.

This small building, location of a one-room schoolhouse, including just a few local students, from grades 1 to 7, was a big task for one teacher. Main event of the year was the Christmas concert giving us an opportunity to show off what we had accomplished in this small learning centre. Toward the end of the evening Santa Claus would appear, Uncle Carl dressed complete with black boots, red suit, and white beard, sliding down a pole from a trap door in the ceiling and attic above, to hand out our presents, with a jolly HO, HO, HO.

Today, Uncle Carl was guiding those two big horses and sleigh, approaching me quickly, between mounds of snow piled high above my head on each side of the narrow country road that led to Rougemont Mountain, snow placed there by ploughs during the night. Seeing Uncle Carl close by, I looked up at him with a serious face, raised my right arm straight out, uttered those infamous words heard somewhere, "Hail Hitler." Propaganda had somehow reached Canada!

"My little prank will amuse him," I thought, but the frown did not leave his brow as he quickly passed me by, with no word, or friendly gesture.

I continued walking home alone, my inseparable cousin John Standish usually walked with me, not today, sick, staying home. My journey, as usual, passed St. Thomas Anglican Church built by our ancestors, a big plaque near the tall bell steeple read 1840. This was now 1943 finding us in the depths of World War Two.

Uncle Carl was not pleased; he soon let me know, but this time with a very serious frown that I had never seen before. The history lesson given that day, not soon forgotten; to Uncle Carl my action was not so simple, ideas expressed to me, hard for a young boy of eight to fully grasp. I was glad uncle did not tell my parents about that silly prank!

Later years gave me understanding why uncle was so upset at my behaviour. Hitler and his fascist regime were teaching eight-year-old children like me a radical idea that they were a Master Race[2]; this taught by Hitler their unquestioned leader. The Aryan race was at the pinnacle of Evolutionary development, superior to any other, ideas being promoted that their race was the result of the "Darwinian Evolutionary Principle[3]" of "Survival of the Fittest[4]," ideas now believed and being taught in classrooms around the world. This relatively new philosophy would give them the right to impose their ideas on the rest of humanity, with justification, regardless of morality, eventually plunging the world in a war for "Survival of the Fittest[5]"

Evolutionary thinking ideas from the past was in direct contrast to the Judeo Christian principles taught in that one-room schoolhouse and in the old Anglican Church, creation taught then, instead of evolution theories taught now. I often looked at the Ten Commandments, the moral standards passed down to me, gracing the wall at the back of that old church. Children received a gold star for memory work placed in a booklet as we progressed with knowledge from a book called the Bible. We memorized verses that included....Love your neighbour as yourself...Thou shall not kill...Honour your father and your mother...be kind one to another...judge not lest ye be judged...learning lessons about compassion and care for the poor...and more...much more. I was not alone in learning this; these Judeo-Christian principles form the foundation for the Western world.

Hitler had achieved absolute control, establishing the Third Reich, which was a tribute to his predecessors, emperors in Europe for the First and Second Holy Roman Empire[6]. These hierarchal systems had governed the world with an iron fist for centuries, but without success to control the British Isles, the last attempt being a vast flotilla of ships known as the "Spanish Armada[7]." This action was a failure!

The Jewish people introduced the world to the concept of Spirit, the supernatural, and the infinite, as opposed to only a finite or natural world. Hitler believed that these ideas had corrupted the world, weakening the human race, and with determination was preparing to eradicate them. A similar policy called Eugenics[8] was justification to exterminate the weak, handicapped, and mentally ill, before and during the Second World War. The world suffered as these ideas of Social Darwinism[9] applied to any nation who dared to oppose, subjecting that nation's people to slavery and oppression, while we witnessed a brief return to barbaric rule, as the historical record of the Roman Empire reveals.

The principles of Evolution and Social Darwinism were admired and well known by the leaders of the Nazi regime, information about their beliefs readily available today.[10] This idea of "Survival of the Fittest[11]" also formed the thinking of autocratic rulers in other nations to produce the many tragedies of the 20th century, ideas that gave us the bloodiest century in history!

History teaches us that an incorrect philosophy can lead to disastrous results. The Philosopher George Santayana[12] said it well, *"Those who do not remember the past are condemned to repeat it."*

It is truly amazing that we continue to teach "Darwinian Evolutionary Principles" and "Survival of the Fittest[13]" in our schools and centres of higher learning. Taught with zeal but not considered "self made" religion. This teaching continues even though there is more than adequate evidence and proof that the whole concept is scientifically untrue. Western civilization with its successful foundation of Judeo-Christian principles slowly changed to a new state supported doctrine.

The Christian philosophy taught to my generation in school and church on the other hand, no longer allowed in our public school systems. This is tragic because some of the very best literature is found in a book called the Bible. These ancient documents complete with many practical lessons now classed as religion only, or creation ideas. The implication is that they are a violation of rights based on the false interpretation of "Separation of Church and State[14]."

Some influential advocates of Darwinism and Atheism even suggest that the teaching of Christian Principles to our children in the home or elsewhere outlawed by the state! This indeed is an old idea; there has already been an attempt by communist countries to control what our children think. This sad fact is also a part of the history of the 20th Century.

The Evolution Theories include a philosophy that is not moral in its implications of "Survival of the Fittest[15]" but it has gained a legal status in our society. History records the tragic result of this philosophy but we seem to choose to ignore history.

Do we need a good lawyer or legal entity brave enough to get this outdated information corrected in court? Do we need a person or group with scientific knowledge to challenge these theories? As clever as the members of the circle of influence who chose to promote these radical new ideas over 150 years ago?

We have come a long way in our organized knowledge since then that proves that these ideas and the philosophy that is applied as a result are scientifically incorrect. History has shown us a perfect example of Cause and Effect[16], the result of this intellectual error.

It is my experience that, *"We Progress when Freedom Reigns."* My ancestor Stephen Bachiler[17] landed in the New World in 1632, told to leave his native England. Having Puritan[18] non-conformist[19] beliefs, this Anglican Church minister wanted to place a greater emphasis on biblical principles that promote freedom. He paid much for his beliefs and persistence!

This new freedom was a dangerous philosophy for king and nobility who exercised control over their people. This minister's belief and that of the Pilgrim Fathers[20], and other leaders that thought like them, eventually formed the Constitution of a new nation that promoted freedom and liberty for all. The nation called United States with a quote by Daniel Webster in January 1830:

"It is, Sir, the people's Constitution, the people's government, made for the people, made by the people, and answerable to the people[21]," with a principle placed on their money reading *"In God We Trust."*

It seems ironic to me that the United States and other nations claim a belief in a Creator God even including it on their currency and in the national anthem that they sing. Academic activity and scientific research use this currency but seldom place "trust" or consider the possibility of a Creator God!

This foundational belief in creation, complete with a Creator, built a strong nation. A thinking process based on

106

Judeo-Christian principles that include creativity and hard work, promoting scientific knowledge, resulting in the development of a myriad of inventions, now gracing the world.

Then there was the "freedom principle," the Afro-American people finally freed from slavery, a people considered lower in a hypothetical "evolution chain" deserving treatment as slaves! These people once freed, in a relatively short period, excelled, even providing a president for that now most powerful nation!

Therefore, I like the statement that Daniel Webster[22], a descendent of Stephen Bachiler[23] once made, *"God grants liberty only to those who love it and are always ready to guard and defend it."*

Are we losing our freedom to think, to defend the truth?

References:-

1. Daniel Webster This page was last modified on 29 July 2014 at 02:28 09August2014/ <http://en.wikipedia.org/wiki/Daniel_Webster>

2. Master Race This page was last modified on 7 August 2014 at 21:18. 09August2014/ <http://en.wikipedia.org/wiki/Master_Race>

3. Darwinism This page was last modified on 29 July 2014 at 03:28. 09August2014/ <http://en.wikipedia.org/wiki/Darwinism>

4. Survival of the Fittest This page was last modified on 29 July 2014 at 16:46 09August2014/ <http://en.wikipedia.org/wiki/Survival_of_the_fittest>

5. Ibid.

6. Holy Roman Empire This page was last modified on 6 August 2014 at 22:02. 09August2014/ <http://en.wikipedia.org/wiki/Holy_Roman_Empire>

7. Spanish Armada This page was last modified on 7 August 2014 at 10:31. 09August2014/ <http://en.wikipedia.org/wiki/Spanish_Armada>

8. Eugenics This page was last modified on 30 July 2014 at 04:01. 09August2014/ < http://en.wikipedia.org/wiki/Eugenics>

9. Social Darwinism This page was last modified on 29 July 2014 at 21:34. 09August2014/ <http://en.wikipedia.org/wiki/Social_Darwinism>

10. From Darwin to Hitler This page was last modified on 1 July 2014 at 01:37 23September2014/ <http://en.wikipedia.org/wiki/From_Darwin_to_Hitler>

11. Ibid.

12. George Santayana This page was last modified on 27 July 2014 at 12:52 09August2014/ <http://en.wikipedia.org/wiki/George_Santayana>

13. Ibid.

14. Separation of Church & State This page was last modified on 8 August 2014 at 07:05. 09August2014/ <http://en.wikipedia.org/wiki/Separation_of_church_and_state>

15. Ibid.

16. Causality This page was last modified on 30 July 2014 at 18:06. 09August2014/ <http://en.wikipedia.org/wiki/Causality>

17. Stephen Bachiler This page was last modified on 14 November 2013 at 21:20. 13August2014/ <http://en.wikipedia.org/wiki/Stephen_Bachiler>

18. Puritan This page was last modified on 13 August 2014 at 08:25. 13August2014/ <http://en.wikipedia.org/wiki/Puritan>

19. Nonconformist This page was last modified on 11 July 2014 at 12:56. 12 September2014/ <http://en.wikipedia.org/wiki/Nonconformist>

20. Pilgrims Fathers This page was last modified on 29 July 2014 at 00:53. 09August2014/ <http://en.wikipedia.org/wiki/Pilgrims_(Plymouth_Colony)>

21. Second Speech on Foote's Resolution, U.S. Senate, 26 Jan. 1830, 12August2014/ <http://izquotes.com/quote/311624>

22. Daniel Webster Quotes 12August2014/ <http://quotes.liberty-tree.ca/quote_blog/Daniel.Webster.Quote.73C7>

23. Ibid.

Chapter Eleven

Denial of Ancient Jewish Historical Records

I still remember Veronica a new student to my boarding school, her arrival at midterm. She soon became the talk of the boys circle. She was beautiful, with flowing blonde hair curling down to her shoulders. The boys would wait anxiously for her to wear one of those frilly "kitten sweaters," giggling with words that came in whispers to each other. Teenage boys not old enough to express feelings in a manner that would please a beautiful girl, awkward boys, trying to impress each other at the expense of that person. Veronica was the talk of the town but she seemed different, very quiet, too shy to talk, keeping to herself even among the other girls. Why was she so different? She seemed so mysterious, while making such an impression on us.

Our Headmaster sensing the problem decided to deal with those boys' obvious ignorance. He took an opportunity to speak to us during one of his classes, most of the older boys being present. We owe a great deal to the Jewish people was his beginning remarks. He then proceeded to explain salient parts of Jewish history, important information that benefited all humanity, the details of this presentation which now eludes me.

Veronica, he continued, is a person that witnessed the murder of her parents by the Nazi regime during World War II, simply because they were Jewish. We need to give her all the respect that anyone deserves, but especially for her.

Did we fully understand, probably not, but from that day on the boys matured, developing an entirely new attitude towards Veronica while insisting on this from all the younger boys as well.

Another contact with Jewish people came when I was very young living on the farm. My mother was a retired teacher having taught in Montréal for several years. She told us about befriending other teachers, especially two teachers with a Jewish German background. "They were not very popular only because of their background," she recounted not swayed by others.

I remember her saying that her Jewish students were most eager to learn, even willing to stay after regular class for additional instruction.

These two Jewish teachers, with family, made many visits to see my mother usually on a Sunday afternoon then living on a farm in Rougemont, Quebec. They would return to Montréal with crates of mother's berries from her prized strawberry patch. They would also bring many gifts including a coat of "many colors" for me, a coat that my parents could not easily afford. It certainly was a fine coat, however I was not pleased, standing out amongst the other students in school; no one had a coat like mine!

Who are these Jews?

Throughout history there is a record of persecution. Why is this happening?

Well, other than for gifting coats of many colours, as I attempt to turn a very serious matter into humour!

111

In order to get an answer to these questions we need to explore the beginning of the oldest civilizations on earth. At this time there are no Jewish people. We need to look at the history of those early people to understand why there is a race of Jews. This required research which is indeed readily available to us in this the "information age." I use the Wikipedia Encyclopaedia on the Internet as an easy source of information. The same data is also available from a large variety of media. The oldest civilization is an area in the Near East. First mentioned is Mesopotamia[1], present day location mainly in Iraq. Migration soon followed to northern India, now known as Pakistan, Europe, Egypt and China, arriving there mainly via waterway always settling close to rivers a source of essential drinking water, food, and transportation.

Mesopotamia means "between two rivers" a Fertile Crescent[2] that extended North West along the shores of the Mediterranean Sea into what is now known as areas of Iran and Turkey. Tigris and Euphrates rivers flowed through this Fertile Crescent[3] south to the Persian Gulf. The people settling there often called Chaldeans[4], a people noted for producing bronze, a reference to the Bronze Age[5]. Many relics found including sailing ships. There is also evidence that Scribes[6] produced the earliest forms of writing.

The first Jews play an important role in Sumeria[7], a part of Mesopotamia.

I may best explain this important role with a personal story. Personal experience is an excellent teacher, sometimes accompanied by pain!

My friend Michael is Jewish; his son just turned thirteen, a celebration when a Jewish boy becomes a man, called Bar Mitzvah[8]. This is my first exposure to a Jewish Synagogue with a Rabbi in charge. Again, I could not contain my curiosity!

My first question; "What are those three big containers located at the front of the synagogue? Some Synagogues have five," the Rabbi started to explain.

"Why not here," I was bold enough to ask.

"These contain the Torah[9] our Holy books of Jewish history," he continued to explain. "They are very expensive because trained scribes meticulously hand copy them insuring absolute accuracy with any new copy. The new copy is never different than the previous one," but I was ready to object.

"How is that possible?"

The Rabbi patiently continued to explain. "Each character, including punctuation has a unique numerical value. The word, sentence, paragraph and page must add up, agreeing with the numerical value of the previous scroll. The Scribe[10] must start over if there is even a minute error. 'Counters' is the name often applied to Scribes. This method, developed ages ago, ensures absolute accuracy, in reproducing copies."

From that valuable experience, I knew there would be very little difference with modern text when compared with the ancient Dead Sea Scrolls[11] found in the Qumran caves. It works, only minor unimportant changes detected. This is an amazing system to ensure accuracy!

The Jewish Race are chosen messengers; recording a meticulous history of our "beginning."

The first Jew, Abram later called Abraham, was a resident of ancient Mesopotamia. Abraham chose to accept the belief of his ancestor Noah, remembering the flood, a brave decision. The people in Mesopotamia living around him chose to worship things that they created rather than the Creator, similar to events before the flood. Abraham left the city of Ur traveling northwest along the Fertile Crescent[12] to find a new land. The word Jew means a believer in the one God, with Abraham the father of all Abrahamic religions. He was a descendant of Shem, Noah's faithful son.

The two main rivers in Mesopotamia[13] are the Tigris and Euphrates empting waters into the Persian Gulf. The main water source for these rivers comes from extensive mountain ranges to the north, part of Turkey. Mount Ararat[14] is found here, the recorded landing location of Noah's Ark[15]. This area is just north of the first known civilization on earth after the Global flood leaving only Noah, his three sons Shem, Ham and Japheth with their wives. A new start in a land called Mesopotamia[16] then their descendants who migrated around the world. Present world population, in spite of disasters, confirm a recent beginning again after the Global Deluge[17].

The original beginning before the flood was obviously different, an event modern science cannot logically deny, just beginning to understand. A recorded time when humans lived much longer protected from destructive sunrays by a canopy above. A time much different with only a small amount of evidence left for us to examine.

We cannot ignore an historical record passed down to us by the Jewish Race. We often choose to ignore it in spite of modern science continuing to confirm its authenticity with new research.

The story in the Garden of Eden is the beginning of the human journey. The first two humans were not faithful to their Creator. Adam and Eve preferred to believe a lie. All down through history we encounter imaginary stories leading the majority[*] of people away from the truth. They too prefer to

---

[*]Note - See contact the author page

believe a lie. The historical records show us the fatal consequences of false devotion. History records events where a Creator had to wipe out complete generations to eliminate the fatal consequences of such behaviour. Then, we also read of men and women who were faithful to the Creator in every generation. They preferred to believe the truth becoming at odds with popular opinion, dangerously going against the majority.

Noah is such a man, faithful to the Creator in spite of derision. The faithful ones endure hardship throughout history, but persist. Noah lived during a very corrupt period amongst people willing to follow an imaginary lifestyle. A Global flood was necessary to wipe out this generation, an action vital for the preservation of humanity. .

Abraham as well was a faithful follower in his generation. He chose the hard road, the truth of his ancestor Noah instead of following the worship of a host of fabricated gods, rather than showing faith in the Creator. All down through history we find such devoted followers.

Moses was also such a man, trustworthy, receiving favour, a descendant of Abraham and Noah. His people originally came from Mesopotamia[18]. Moses, highly educated, living forty years in Egypt, educated in the best schools, in the Royal Courts of Egypt[19]. He then received forty years of practical experience with knowledge living with descendants of Abraham. With this knowledge and experience, Moses led his people out of a nation with pagan worship. Moses no doubt receives an education in the "art of writing" in the courts of Egypt. He was best suited to provide an historical record of creation, the Torah[20], the first five books of the Bible. This was now a written record rather than "oral tradition[21]" passed down from generation to generation. Moses was only a few generations away from Adam at the beginning. These early writings inspired by a belief in a Creator. The Egyptian people, highly developed in science with material comforts chose to worship the cosmos, sun, moon, and stars, rather than being faithful to a Creator.

Today, we find ourselves in a world of wild suggestions coming from previous generations. These imagining's, seldom questioned, theoretical, without supporting evidence, instilled in the minds of our children.

Do we find a similar teaching in many evangelical and theological circles in addition to the Bible trying to combine evolution with creation? We often hear the theory that we evolved over millions of years presented by every form of media with little or no contrary opinion. Is that Creator Spirit looking for faithful followers today like Abraham? Are men and women willing to discover truth in spite of any derision or adverse action received from a majority viewpoint?

Discovery is readily available today with so much information available with a minimal amount of effort. Today, we need not visit a library, but have a wealth of information available from the computer screen, in our homes, regardless of any remote location. Many choose to believe theory, imaginary ideas, and century's old rather than studying recorded history.

Does our age as in the past, seem to place more emphasis on nature, claiming all evolved from it rather than even considering the possibility of a Creative Spirit?

Therefore, what valuable contribution did the Jewish race give to our world?

They give us a very accurate account of history preserved down through the ages to the present day. If we consider scientific knowledge as well, we find a large percentage of Nobel Prize winners come from the Jewish race[22]!

Did that Headmaster give me some homework after his speech to the senior boys?

I guess he did!

References:-

1. Mesopotamia This page was last modified on 29 July 2014 at 02:20. 09August2014/ <http://en.wikipedia.org/wiki/Mesopotamia>

2. Fertile Crescent This page was last modified on 13 July 2014 at 09:32 09August2014/ <http://en.wikipedia.org/wiki/Fertile_Crescent>

3. Ibid.

4. Chaldeans This page was last modified on 30 July 2014 at 08:01. 09August2014/ <http://en.wikipedia.org/wiki/Chaldea>

5. Bronze Age This page was last modified on 6 August 2014 at 20:39. 09August2014/ <http://en.wikipedia.org/wiki/Bronze_Age>

6. Scribes, This page was last modified on 7 July 2014 at 11:09. 09August2014/ <http://en.wikipedia.org/wiki/Scribe>

7. Sumeria This page was last modified on 20 February 2014 at 05:48. 09August2014/ <http://en.wikipedia.org/wiki/Sumeria>

8. Bar Mitzvah This page was last modified on 4 August 2014 at 15:15. 09August2014/ <http://en.wikipedia.org/wiki/Bar_and_Bat_Mitzvah>

9. Torah This page was last modified on 5 August 2014 at 06:48. 09August2014/ <http://en.wikipedia.org/wiki/Torah>

10. Ibid.

11. Dead Sea Scrolls This page was last modified on 9 August 2014 at 03:56. 09August2014/ <http://en.wikipedia.org/wiki/Dead_Sea_scrolls>

12. Ibid.

13. Ibid.

14. Mount Ararat This page was last modified on 6 July 2014 at 16:28. 09August2014/ <http://en.wikipedia.org/wiki/Mount_Ararat>

15. Noah's Ark Found in Turkey? Published April 28, 2010, <http://news.nationalgeographic.com/news/2010/04/100428-noahs-ark-found-in-turkey-science-religion-culture/>

16. Ibid.

17. Deluge, This page was last modified on 27 May 2014 at 03:22. 09August2014/ <http://en.wikipedia.org/wiki/Deluge>

18. Ibid.

19. Moses (Royal Courts of Egypt) This page was last modified on 9 August 2014 at 14:14. 09August2014/ <http://en.wikipedia.org/wiki/Moses>

20. Ibid.

21. Oral Tradition This page was last modified on 4 February 2014 at 08:18. 09August2014/ <http://en.wikipedia.org/wiki/Oral_tradition>

22. Jewish Nobel Prize Winners This page was last modified on 23 July 2014 at 02:41. 09August2014/ <http://en.wikipedia.org/wiki/List_of_Jewish_Nobel_laureates>

Chapter Twelve

Connecting Natural Evidence to History

Flames rose in an irregular pattern from those towers, both day and night, emitting a distinctive odour, a smell soon forgotten, with time. This was what we witnessed daily as we looked out dormitory windows.

I was at boarding school, for five years, except for a few weeks at home each year. This was residence for several boys and girls, on the shores of the St. Lawrence River. Pointe aux Trembles, a small town at the eastern most end of Montréal Island was the location of this private school. The front of the school faced the River, a source of many large boats plying their way along this famous waterway. The two-way traffic only ceased during the colder periods of winter, icebreakers opening the channels during less severe winter. A constant flow of water usually kept this waterway open, fresh water from sources many miles west from the Great Lakes, largest freshwater reserve in the world, flowing over mighty Niagara Falls on its way to the Atlantic Ocean. The traffic on the St Lawrence often included large oil barges headed for the Montréal East refineries. A large

Empress ship would halt any activity at the boarding school when spotted headed up river to the Port of Montreal.

The most fascinating impression for me and most of the other students were those flames always glowing nearby, a subject of much discussion. These curiosities finally lead to a most interesting visit to many refineries.

I recall all the tanks filled with a variety of oil products connected by an extensive display of pipes, valves, gauges, and gears. We received a detailed description of how a thick oily substance was refined into a variety of products by Cracking Tower[1]. Flames at the top of these huge structures a result of heating crude oil, a process used to extract a wide variety of chemicals.

One of the students, maybe me, asked the inevitable question. What produced all this oil, or some similar direct question?

I still remember that the answer was very small fish sinking to the bottom of the ocean with one drop of oil in each, produced this oil "millions of years ago." I remember questioning that answer that only dead fish sinking to the bottom of the ocean formed the oil that we use today! What made the oil that we find on land?

There were no further questions, that I can recall readily accepting that answer, coming from an official of a major oil company.

Today, any mention of "millions of years ago" raises one of my red flags, a conscious result of personal research that contradicts such a suggestion. This is my additional reasoning; stating that an event took place a very long time ago is a convenient way of not saying "I do not know." This phrase of millions often loosely applies to dinosaur bones as well as the production of oil, a statement often heard of a distant past existence during some ancient period. This presents a problem; we have no historical record of any event beyond a few thousand years past. Let me explain further; starting with dinosaur bones, then I will provide a logical explanation for the production of the many oil products!

The design of our living planet includes many cycles[2] to keep it alive and renewed. This part of nature designed to consume bones complete with various items of garbage. This action readily observed in a garden compost heap, discarded material turned into top soil in just a few years. All material, including bones becomes a feast for beetles and insects, a natural cycle.

We too are part of that cycle of being born and die usually within less than one hundred years. I would rather not explain what happens to our remains in a relatively short period. If this were not part of nature, bones with discarded material from previous generations would be our destiny. An impossible situation if these cycles ever stopped working, as the "deep time[3]" concept developed by Geologist James Hutton in the 18th century, seems to suggest. This is not such a radical conclusion arrived at with only my research as the following will explain.

Mary Highby Schweitzer[4] a palaeontologist at North Carolina State University made a most interesting discovery just a few years ago. In examining the insides of a Tyrannosaurus skeleton from Hill Creek, Montana, she found soft tissue, collagen[5], from a dinosaur, an accepted 68 million-year-old fossil bone. The discovery alarmed her; knowledge indicates that soft tissue does not last for such a length of time. A strong attempt to disprove her research soon developed but without success. Her discovery finally accepted as legitimate.

However, what followed this acceptance was most interesting, showing an obvious bias, only wanting to believe production in a distant past. Instead of considering that the dinosaur must have lived recently there developed another strategy, a concerted effort to find how soft tissue could last for such a long period!

Is it possible that modern research refuses to consider anything that will alter a belief in Evolution with a connection to "Deep Time[6]?" Is this a consistent effort to dismiss any possible connection to recorded history?

When we hear "millions of years ago," concerning fossil fuels we need to deal with another somewhat different dilemma. There are volumes of methane gasses[7] stored with the oil mixtures, held in chambers only a relatively short distance below the surface. I can explain my concern using a very simple experiment familiar to most of us involving balloons.

A helium balloon filled with gas lighter than air, similar to methane, causes a balloon to rise stopping on the ceiling. It will stay there for possibly a week, you will find it then on the floor, all helium escaped. To suggest that methane gasses remain below ground under extreme pressure for multiple millions of years seems a most unlikely conclusion. There are also unconfirmed reports that methane gases are constantly escaping at various locations around the world affecting navigation as well as our climate. Basic reasoning suggests that dinosaur bones could be recent along with the production of fossil fuels and natural gasses.

This seems to call for a new look at the natural world that surrounds us based on current research rather than just concluding that all happened "millions of years ago."

My research continued; let me start with a brief understanding of the nature of dinosaurs because they help prove some very important past events. One of the first discoveries of such a creature at least in the recent past was by William Buckland[8] from England, finding huge bones in 1824 in Stonefields near Oxford. He chose to call it Megalosaurus[9] identified even then as a type of prehistoric reptile. Sir Richard Owen[10] named it dinosaur[11] in 1841 meaning Terrible Reptile or Fearfully Great Reptile. These reptile[12] remains found today around the world, usually with only a large bone exposed, seldom more. These bones then used to construct the whole reptile. These creatures come in a variety of sizes and shapes but are usually massive; the plant-eating types can weigh several tons.

China calls them dragons[13] using a replica in their various ceremonies. This country and people recognized as one of the oldest civilizations next to Mesopotamia[14] known as the oldest. The Chinese maintain old traditions, the design of dragons, possibly recalling a recent past when huge reptiles roamed their land. A part of life from the past, giving today's Chinese knowledge to reproduce a likeness of these massive dragons.

Reptiles[15], sometimes called dragons[16], along with many other names, demonstrate a unique feature unlike any other creature; shedding their skin while continuing to grow, much like the present day snake. The present reptiles limited growth only a factor of life span, access to vegetation, or desirable environment.

Mary Highby Schweitzer[17], the palaeontologist mentioned earlier stated in a paper,

"As far as we know, the way the lung tissue functioned, the way the haemoglobin functioned, was designed for an atmosphere that's very different than today's."[18]

I too observe a problem; the very small nostrils of these huge reptiles do not give them enough air to support such a huge body. Even smaller animals today, such as the horse demonstrate big flailing nostrils to feed their bodies with air, especially if stressed during a race.

"An asteroid striking our planet 'millions of years ago' wiped out all these reptiles" science informs us. If that were so, logic tells me that lesser life forms would also die after such a natural disaster. I disagree that the dinosaur is extinct because I discovered many still living which I explain in a subsequent chapter!

Then, is there adequate evidence to suggest that these reptiles lived with man in the recent past?

We find reference to these reptiles in many ancient civilizations as well as the historical record as stated in the following descriptions:

Mesopotamia, considered the oldest civilization, location of a new beginning for humanity after the flood, in the Fertile Crescent just south of where Noah's Ark landed. Artefacts discovered there such as pottery, display drawings of these huge creatures.

China previously mentioned, another ancient civilization, use a replica of these creatures in their ceremonies.

Egypt, part of the Fertile Crescent, also an ancient civilization prominently displays these creatures in drawings, as well as in many replicas.

125

Jewish History in a book called Job, part of the Biblical historical record, gives an excellent description of a behemoth meaning large beast. This is a detailed word picture found in Job chapter 40: verses15 to 19, a description very close to what we now know as dinosaurs:

"Look at the behemoth, which I made along with you and which feeds on grass like an ox. What strength he has in his loins, what power in the muscles of his belly! Its tail sways like a cedar; the sinews of his thighs are close-knit. Its bones are tubes of bronze, his limbs like rods of iron. It ranks first among the works of God, yet its maker can approach him with his sword."[19]

Is it time for us to review an undisputed aging method that states millions when evidence reveals thousands of years for the existence of these huge reptiles?

I now turn my comments to the production of fossil fuels. When we examine these carbon fuels, we also hear production is in "millions of years ago." Regardless of the name, fossil fuels, crude oil, petroleum, natural gas, or just oil, the common name is "carbon[20]." For example, an internet search on one of these carbons reveals the following;

*"**Natural gas** is a fossil fuel formed when layers of buried plants and animals are exposed to intense heat and pressure over thousands of years. The energy that the plants originally obtained from the sun is stored in the form of chemical bonds in natural gas. Natural gas is a non-renewable resource because it cannot be replenished on a human time frame."[21]*

126

All life forms on our planet depend on carbon[22]. It forms a large percentage of our body, arriving there when we consume it from plants and animals that have previously done so. It is very versatile readily combining with other "elements" to form compounds. The largest source existing today is in our forests, trees, commonly called wood. Trees provide a very large quantity of carbon above ground as well as an extensive quantity in root systems under ground. Trees along with all vegetation are a result of our sun emitting rays to convert carbon dioxide into carbon through a process called photosynthesis. The by-product is oxygen essential for all life forms.

We have much more scientific knowledge about this most important product than in the past and I include the following brief description for some of them:

***Charcoal,*[23]** is cooked wood in the absence of oxygen, turning it into a soft form of coal.

***Coal,*[24]** is similar to charcoal in various forms depending on the conditions and how much heat and pressure were present.

***Fossil Fuels,*[25]** a carbon product, wood, plant, or animal material, subjected to extreme heat and pressure. We can produce it artificially but this is very costly, much cheaper to extract it from beneath the ground already produced there under ideal conditions.

***Diamonds,*[26]** other carbon products are the hardest known natural substance. Diamonds found in the bowels of the earth now produced artificially. For example, an animal or pet, even a human, reduced to a diamond in about two weeks, this is now offered by an enterprising company![27]

However, when I do an internet search on diamonds, I read it takes billions of years to produce them under conditions of intense heat and pressure! All this evidence challenges the accepted idea that oil, coal, natural gas, and diamond require a very long period for their production.[28]

Science alone is inadequate in explaining all the actions of the past without accepting a historical record as it chose to do for centuries. The historical record provides knowledge about human progress, actions of humanity over time, within nations. I combine science with history with a result rarely presented to us.

What can we discover when we combine science with history! The book of Genesis describes the beginning of our Natural World. Jews, Christians, and Muslims, know this story. A meticulous method of accuracy preserved the story by the Jewish race.

The story starts with a Spirit hovering over our planet Earth much like the other dead planets in our solar system. The major difference, it was covered in water.

This Spirit proceeds to transform this water filled planet into one that supports life forms complete with vegetation for food in a logical series of events. Rotation in the presence of our sun was one of the first actions giving us day, night, and a time reference. The other rotation around that light source, the sun, we call a year.

The story proceeds to describe a logical sequence of events for five days, forming an ideal place for humans and all life forms.

On the second day that Spirit divides the water from the surface of the earth moving it to form a water canopy in the heaven above later called the firmament. The story goes on to state that this canopy of water was broken descending back to Earth about 2000 years later. The Antediluvian[29] period is from the original creation up to the Global flood. The falling of all this water turned a previously very large land surface into one now largely covered in water.

What would be the natural condition of our planet during the Antediluvian period? I submit the following condition on a planet with a much larger land area:

Sun, rays required for photosynthesis filtered, removing undesirable radiation effects coming from both sun and outer space. This removal of detrimental rays provides a healthier environment with longer life spans for all life forms and plants.

Temperature of our planet is uniform producing a greenhouse effect. This is an ideal condition for growth of an abundance of vegetation for food and oxygen for all life forms, while they produce carbon dioxide in a synergistic[30] relationship.

Atmosphere, a tropical paradise with elevated quantities of oxygen and carbon dioxide gases a natural result, with carbon dioxide for massive plant growth, oxygen for an abundance of large life forms including dinosaurs.

Longevity, as reported in the historical record, humans living for hundreds of years. Reptiles, now called Dinosaurs, becoming huge because we know they never stop growing. These were ideal conditions for reptiles with an abundance of food and elevated amounts of oxygen for their tiny nostrils.

However this idyllic period ended when the canopy descended in a deluge of water for 40 day and nights, with previous extensive landmass now largely covered by water. The massive quantity of vegetation, animal and marine life uprooted, transported to new locations, buried by surging walls of water with sediment changing the landscape everywhere.

A limited land mass and extensive water reservoir now buries massive volumes of versatile carbon under conditions of extreme heat and pressure. Ideal conditions are now present for the manufacture of coal, crude oil, natural gasses and diamonds[31]. This is nature's action on carbon[32] producing a wide variety of hydrocarbons[33] that we now know how to produce artificially using a similar application of intense heat and pressure. The bowels of earth in every location bury this carbon. These historical events are from a few thousand, not millions of years ago, but we choose to ignore the evidence!

Therefore, how can we continue to ignore the Genesis account, description of a beginning, conditions under a water canopy, effects on massive amounts of carbon by a Global flood, and the resulting abundance of evidence? Can we find any other reason for occurrences without simply adding that it all happened "millions of years ago?"

References:-

1. Cracking Tower This page was last modified on 4 August 2014 at 14:50. 09August2014/ <http://en.wikipedia.org/wiki/Catalytic_cracking>

2. List of Cycles This page was last modified on 11 July 2014 at 00:29 09August2014/ <http://en.wikipedia.org/wiki/List_of_cycles>

3. Deep Time This page was last modified on 7 June 2014 at 23:27. 10September2014 <http://en.wikipedia.org/wiki/Deep_time>

4. Mary Highby Schweitzer This page was last modified on 18 May 2014 at 06:50. 09August2014/ <http://en.wikipedia.org/wiki/Mary_Higby_Schweitzer>

5. Collagen This page was last modified on 6 August 2014 at 20:48. 09August2014/ <http://en.wikipedia.org/wiki/Collagen>

6. Ibid.

7. Methane Gas This page was last modified on 28 July 2014 at 14:50. 09August2014/ <http://en.wikipedia.org/wiki/Methane>

8. William Buckland This page was last modified on 25 July 2014 at 09:03. 09August2014/ <http://en.wikipedia.org/wiki/William_Buckland>

9. Megalosaurus This page was last modified on 13 July 2014 at 17:32. 09August2014/ <http://en.wikipedia.org/wiki/Megalosaurus>

10. Sir Richard Owen This page was last modified on 30 June 2014 at 13:32. 09August2014/ <http://en.wikipedia.org/wiki/Richard_Owen>

11. Dinosaur This page was last modified on 9 August 2014 at 10:11. 09August2014/<http://en.wikipedia.org/wiki/Dinosaur>

12. Reptiles This page was last modified on 4 August 2014 at 04:29 09August2014/ http://en.wikipedia.org/wiki/Reptile

13. Dragons This page was last modified on 6 August 2014 at 05:41 09August2014/ <http://en.wikipedia.org/wiki/Dragon>

14. Mesopotamia This page was last modified on 29 July 2014 at 02:20. 09August2014/<http://en.wikipedia.org/wiki/Mesopotamia>

15. Ibid.

16. Ibid

17. Ibid.

18. Article called Schweitzer's Dangerous Discovery from Discover Magazine, April 2006 Issue

19. Job 40: 15 to 19 New International Versions

20. Carbon This page was last modified on 5 August 2014 at 07:58. 09August2014/ <http://en.wikipedia.org/wiki/Carbon>

21. Article obtained Nov. 10, 2013 from United States Environmental Agency.

22. Ibid.

23. Charcoal This page was last modified on 7 August 2014 at 09:36. 09August2014/ <http://en.wikipedia.org/wiki/Charcoal>

24. Coal This page was last modified on 9 August 2014 at 06:49. 09August2014/ <http://en.wikipedia.org/wiki/Coal>

25. Fossil Fuels This page was last modified on 24 July 2014 at 04:17. 09August2014/ <http://en.wikipedia.org/wiki/Fossil_fuel>

26. Diamond This page was last modified on 5 August 2014 at 20:54. 09August2014/ <http://en.wikipedia.org/wiki/Diamond>

27. Memorable Diamonds<http://www.memorial-diamonds.com/

28. Ibid.

29. Antediluvian This page was last modified on 2 August 2014 at 20:00. 09September2014/ <http://en.wikipedia.org/wiki/Antediluvian>

30. Synergistic This page was last modified on 24 August 2014 at 10:03. 09September2014/ <http://en.wikipedia.org/wiki/Synergy>

31. Ibid.

32. Ibid.

33. Hydrocarbons This page was last modified on 11 September 2014 at 01:55. 11September2014/ <http://en.wikipedia.org/wiki/Hydrocarbon>

Chapter Thirteen

Connecting Supernatural Evidence to History

"Working with you was a pleasure; you were consistent during all our sessions together," my lawyer commented as I prepared to leave his office. This was the last expensive meeting with a family lawyer, trying desperately to find a way to keep my family together to avoid a divorce. Being quick to respond to my lawyer's comment I said, "What do you mean by consistent?"

He proceeded to explain a technique used to determine the truth. "Ask a similar question in slightly different ways, then see if the answer is the same," he further explained. "If these answers are consistent during several meetings one may conclude that the truth is being revealed." Quite logical, a lesson I did not soon forget. That comment from a lawyer, unfortunately, did not prevent a subsequent family divorce.

This was a time in my life when I experienced breakup of the family involving innocent children. Confusion and stress finally led to my depression. This action involves lawyers with a court system removing a family unit, then some family assets, if not all of them. My lawyer had trouble convincing me it was not about right or wrong rather what was legal. He finally convinced me to accept my legal fate. My family was now in the hands of a system that I felt was more interested in legality than the future condition of innocent children. That action took responsibility away from me as head of a family.

I questioned belief in a Creator that would allow such pain and sorrow. My subsequent journey of discovery led me to certain conclusions. My lawyer, unaware of his participation, helped in my research. I learned much from that lawyer about life. Sessions that were very expensive are part of the cost of divorce. The love bond was broken between children and parents, causing damage. This certainly was not the fault of my children, but rather the actions of their parents.

As head of a company, you learn there comes a time when you have to "fire," let someone go. The penalty for that employee is removal of salary and benefits. The bond of trust is broken between employer and employee making it necessary to remove access to protect the company. In the case of the creation story, a trust with love bond is broken between Creator and the created. Adam and Eve believed a lie from the serpent rather than the truth given them by the Creator, deciding to follow a different path that led to death. This disobedience by Adam and Eve also caused their removal from the Garden of Eden including access to directions from their Creator. A relationship was broken with damage to the plan of creation including the humans given the responsibility to manage it. All these events are described in the historical record, the biblical book of Genesis.

This action by Adam and Eve is similar to a parental breakup with children as previously described but in this case the children, the created, broke with their Creator. In spite of this disobedience, history records how the Spirit provides a way for us to return to a relationship once more. This was the original plan of creation, an on going relationship with the Creator.

I again considered what I learned from that lawyer when he stated that consistency[1] indicates truth! We may have trouble considering anything spiritual as it is so foreign to our physical existence or way of thinking. It may be easier to consider an Alien visiting our planet in a physical body via a flying saucer. There is no evidence of such an arrival so we may conclude that our government is hiding flying saucers with little green men. A desire to connect with another world similar to ours is a strong one!

However, when we consider another world I suggest that we need to think of a spiritual dimension rather than only the finite world of our existence. I listened to a speech recently, which reasoned that if we could possibly travel at the speed of light[2] about 300,000 kilometres **per second** (186,000 miles **per second**) it would take us over 4 years to reach the closest star, Proxima Centauri[3]. However, we must understand that the fastest that we can now travel is much less than 160,934 kilometres **per hour** (100,000 miles **per hour**) only a "small fraction" of the speed of light![4]

We attempt to describe travel in a universe with distances and sizes when the word "immense" is inadequate by using "day" and "year" as a measure, the only means of measuring time available to us. These measures come from our earthly experience, day being one rotation of our planet and year is roughly 365 days of rotation around our sun. The universe seems to demand a much different "time scale[5]" to measure it. There is no evidence that an Alien could visit or is able to do so in a "bodily form" similar to ours. It seems reasonable to me that any Alien visit would involve a very different "bodily form." The historical record in Genesis describes such a visit as a spiritual being rather than physical. The story goes on to describe how that Spirit converted a pre-existing dead planet into one supporting life. A dead planet is then converted to one designed for our bodily existence with our internal Spirit able to connect with that Creative Spirit. However, we destroyed that connection with the Spirit when we decided to follow our own way in defiance to that plan for us.

In the creation story that Creative Spirit uses "words" to create! Does the story follow a path of consistency[6] that I learned from my lawyer?

The Biblical story provides consistency[7] with many predictions followed by fulfillment later. The book of Genesis describes a sequence of events that are undeniable. The Biblical story goes on to provide details of how the Spirit decides to visit us in a "bodily form," like us. His name is Jesus, which means Saviour in the Jewish language. The ultimate purpose for his visit is to renew a relationship with us, broken when we believed a lie at the beginning of time. Death is the ultimate penalty for destroying a perfect creation, one described as "very good[8]." The story continues telling us that Jesus took the penalty of death for us!

The following are a few highlights that my lawyer had referred to that provide evidence for consistency[9]. The first references are predictions, often called prophecy, from the Biblical Old Testament. The second reference is evidence, often called fulfillment from the Biblical New Testament. There are

"60 Prophecies[10]", an extensive list in the internet site under that name. Jesus the Spirit that became "bodily flesh" as described in the historical record, the Bible, fulfilled all these prophecies! A few of these predictions with subsequent evidence are listed as follows:

His Birth is from the seed of a woman, of a virgin, in the family line of Abraham, Jacob, Jesse and King David. He is born in the town of Bethlehem in a lowly manger.

Prediction: *Genesis 3:15, Jeremiah 23:5, Micah 5:2*

Evidence: *Galatians 4:4, Luke 3:31, Matthew 2:1-6*

His Nature- He pre-existed creation called Immanuel (God with us) and anointed by the Spirit.

Prediction: *Micah 5:2, Isaiah 7:14, Isaiah 11:2*

Evidence: *1 Peter 1:20, Matthew 1:22-23, Matthew 3:16-17*

His Teaching, a messenger precedes him. (John the Baptist) Jesus provides a ministry of miracles. He entered Jerusalem on a donkey.

Prediction: *Isaiah 40:3, Isaiah 35:5-6, Zechariah 9:9*

Evidence: *Matthew 3:1-3, Matthew 9:35; 11:4, Matthew 21:1-7*

His Betrayal, sold for 30 pieces of silver by a friend named Judas who betrays Him. False witnesses then accused Him.

Prediction: *Psalms 41:9, Zechariah 11:12, Psalms 35:11*

Evidence: *John 13:18-27, Matthew 26:14-15, Matthew 26:59-61*

His Crucifixion commits himself to God. They pierce His side then bury Him in a rich man's tomb.

Prediction: *Psalms 31:5, Zechariah 12:10, Isaiah 3:9*

Evidence: *Luke23:46, John19:34+37, Matthew 27:57-60*

His Resurrection and Ascension rose from the dead, the begotten as Son of God. He ascended to God.

Prediction: *Psalms 16:8-11, Psalms 2:7, Psalms 68:18*
Evidence: *Acts 2:24-31, Acts13:32-35 Ephesians 1:3+13*

Jesus performed many miracles, supernatural acts recorded in the New Testament, healing the sick, restoring the blind to sight, bringing the dead back to life, to mention a few. These are supernatural actions performed with just "words." This may be hard to believe but it is consistent with the historical record in Genesis when all creation happened with "words" from that Creative Spirit.

There are eyewitnesses to these supernatural events by many, including four close followers, Matthew, Mark, Luke, and John, who wrote it down for us. This action with "words" only is described by John, one of those close followers of Jesus, as follows:

"In the beginning was the Word, and the Word was with God, and the Word was God."[11]

Then later John wrote:

"The Word became flesh and lived for a while among us. We have seen his glory, the glory of the one and only Son, who came from the Father, full of grace and truth."[12]

The first recorded supernatural action performed by Jesus was with a "word" only when He changes water into the best wine. This compares to an action at the "beginning" recorded in the book of Genesis. In six days the supernatural actions of a "word," by the Creative Spirit changes a water covered "dead planet" into a "living one."

Did this very first supernatural action by the Spirit that became flesh involving water, a second time, provides evidence of Jesus power as the Creator of Genesis? Does it also prove His mastery over nature in this our natural world? Do we discover consistency in all these actions?

As explained before His main purpose in becoming one of us was to be a Saviour taking the penalty of death in our place. This resulted in His death on a cross, an action to benefit those who believe in Him, giving us a promise, as follows:

"For God so loved the world that he gave his one and only son, that whoever believes in him shall not perish but have eternal life"[13]

Jesus also said he is preparing a place for those who earnestly believe in Him!

"In my Father's house are many rooms; if it were not so, I would have told you. I am going there to prepare a place for you." [14]

He promised that if we believed in him we would be free to discover the truth[15] and that he would return again[16].

Therefore, we can consider all the consistent evidence provided for us or we can ignore it. Adam and Eve received this freedom to choose at the very beginning; their decision to accept or ignore the Creator was their free choice. This is the same freedom, a free choice given to each one of us, today! We have that same freedom to accept or reject a Creator.

The choices we make determine our destiny; choose well my friends!

References:-

(All biblical references are from the New International Version)

1. Consistency This page was last modified on 22 July 2014 at 20:00. 16August2014/ <http://en.wikipedia.org/ wiki/Consistency>

2. Speed of Light This page was last modified on 8 August 2014 at 00:52 16August2014/ <http://en.wikipedia.org/ wiki/Speed_of_light>

3. Proxima Centauri This page was last modified on 16 August 2014 at 08:06. 15August2014/ <http://en.wikipedia.org/wiki/Proxima_Centauri>

4. Ibid.

5. Time Scale This page was last modified on 21 March 2013 at 05:28. 10September2014/ <http://en.wikipedia.org/ wiki/Time_scale>

6. Ibid.

7. Ibid.

8. Very Good, Genesis 1:31 "God saw all that he had made, and it was very good. And there was evening, and there was morning--the sixth day."

9. Ibid.

10. 60 Prophecies, <http://www.bible.ca/b-prophecy-60.htm>
15August2014/

11. John 1:1

12. John 1:14

13. John 3:16

14. John 14: 2

15. John 8:32

16. Mark 13:26

Chapter Fourteen

Dear Charles Darwin

Charles Darwin died on April 19, 1882, buried in Westminster Abbey apparently after a very elaborate funeral. If I could write him a letter today this is what I may include:

Charles Darwin
Westminster Abbey
London, England

Dear Charles,

This is a rather informal letter. I trust you will accept it in good faith as I have heard so much about you, feeling that I know you personally.

Your contribution to science is significant, along with others, in starting a whole generation of thinkers, part of the "Age of Enlightenment."[1] Since then science has contributed much to our knowledge about the natural world. I would like to inform you concerning some of the developments of the last two hundred years. This involves the theories developed by you, James Hutton[2], Charles Lyell[3], and others.

There have been numerous attempts to prove that we evolved from chemicals millions of years ago. All these experiments have failed.

Of the 7 billion humans now on our globe with billions coming before them, each one has arrived from life. It seems that your contemporary Louis Pasteur[4] is correct "Life only comes from life" based on the Law of Biogenesis[5], not from a laboratory or nature. Your friend Thomas Huxley[6] tried to promote an opposite theory, Abiogenesis[7] that life can come from nature. However, there is no supporting evidence for that theory or from nature!

We now have very powerful microscopes to disprove the idea that a single cell was just a "blob" of inert chemicals. We have discovered that each cell in our body has multiple thousands of instructions called DNA[8] necessary to create only that "kind" of living being. Every human, along with all living creatures has a unique set of instructions.

However, you certainly were right about the "eye," much too complex to evolve from inert chemicals. We have a better understand of each organ in our body as being complex beyond imagination, all working in an intricate plan to make the human body function. These combinations of necessary organs we realize are very complex. With our very powerful microscopes, we even discover this complexity in the minute bacterial world.

Exploration is now a reality to various parts of the solar system including parts of the cosmos. Intelligent design is increasingly evident. We even dream of travel to Mars. Our probes of that planet indicate no oxygen to breathe, daily temperatures vary from high degrees during the day to extremely low degrees at night, no carbon based fuel for heating, sun very dim, no food to eat, no water to wash or drink, and violent storms to survive. It appears to be quite a challenge, giving us a renewed appreciation of the magnificent planet designed for us!

It seems that the more knowledge we acquire continues to point to the intelligent design of planet earth rather than unguided processes. This includes a plan for a living planet, all life forms, vegetation, within a planned solar system. Some of us are reconsidering the original historical account of Genesis for answers, rejected during the "Age of Enlightenment[9]" in favour of just natural science.

Your friend and colleague William Whewell[10] promoted the idea of Inductive Reasoning[11] instead of Empirical Evidence[12]. Inductive Reasoning is basically using your mind to come up with imaginary ideas then testing them to see if they are true. The progress of recent knowledge has disproven theories based on this method proposed by your friends and colleagues. I describe just twelve of these incorrect theories, leading to false premises:

1ˢᵗ False Premise-Ignore Jewish historical records

The ancient Jews devised a very clever mathematical method to ensure that their records were an accurate but easy to understand description of beginning events.

These chosen messengers preserved the original text by using this method. We cannot ignore an historical record in any scientific research if we want to discover the truth.

2ⁿᵈ Deep Time, this planet Earth is millions of years old.

In 1972 we invented the Atomic Clock[13], a very precise time piece so we now know that the rotation of our planet slows at an average rate of about 1 second every 500 days (one and one half years) called Leap Seconds.[14] This fact is a major blow to your evolution theories because there are only about 86,400 seconds in a day (60 seconds times 60 minutes times 24 hours). If we choose to do the math our "living planet," therefore, cannot possibly be millions or billions of years old!

3rd *There was no Global flood*

The fossil record indicates instant burial with oxygen removed so that the source could not decay being available for us to discover.

We now know that very high heat and extreme pressure along with water is required to produce hydrocarbons such as natural gas, coal, oil, and diamonds. This flood action is evident all around the world with the deposit and burial of a variety of hydrocarbons.[15]

4th *Uniformitarianism (Uniform Action)*

For the last 60 years or so we have been able to descend to the depths of our oceans and discover thousands of extinct volcanoes, evidence of an event of immense proportion that happened in the recent past. We are able to see on a screen in our homes what even one earthquake does to water rising up onto land, moving large objects and buildings like pieces of driftwood. We now understand what thousands of volcanoes can do to the vast ocean waters and adjacent land. It would obviously affect the land for thousands of miles, affecting mountains, valleys, lakes, and creating massive depressions.

Uniform action is not evident anywhere, even in our life time regardless of considering any ages in millions or billions of years.

5th Names & Ages of Sedimentary layers & millions of years

James Hutton[16] suggested in the late 1700's that no catastrophic event ever occurred basing his idea on observations at his farm. He proposed that sedimentary layers were laid there over millions of years. This was his suggestion in spite of no erosion between layers during this vast period. Charles Lyell[17], your associate, accepted this earlier suggestion by adding descriptive titles to some of the sedimentary layers while heavily promoting the idea. These theories are not a product of empirical science.[18] However, scientists even today use these theoretical ages as a premise for their research!

A noted Geologist sums it up very well for us:

"Eighty to eighty-five percent of earth's land surface does not have even 3 geological periods appearing in 'correct' consecutive order ... it becomes an overall exercise of gargantuan special pleading and imagination for the evolutionary-uniformitarian paradigm to maintain that there ever were geologic periods."[19]

John Woodmorappe, Geologist[20]

6th Natural Selection

You reasoned that micro changes, known today as mutations, eventually evolved by natural selection into macro changes to produce a new different kind of species. We now know that intelligent design includes mutations[21] but never a total change to a different "kind" that requires an entirely different plan, and instructions.

7th Survival of the fittest

The evolution philosophy unlike other religious philosophies does not include any code of conduct toward our fellow man or other creatures on earth. This philosophy states that the fittest will survive because they are at the top of the evolutionary chain. Dictators and some religions still use this philosophy today as a reason to enslave or even murder others, sometimes at random.

Recent history has proven the folly of this type of thinking with very destructive wars, killing millions, and affecting all humanity, even justifying acts of terror.

8th Life evolves from nonlife.

We cannot produce life from non-life, or chemicals no matter how hard we have tried, it requires life to produce life and we see that action all around us. Any other suggestion violates the Law of Biogenesis[22] previously explained.

9th Nature goes from simple to complex

The natural world does not go from simple to complex but rather complex to simple. Everything has a beginning and an end. We only have to look at our own life cycle. We are born and will eventually die. This conforms to our Laws of Thermodynamics.[23]

10th There is no intelligent design.

We have discovered what we call DNA[24] now realizing that a single cell is not just a blob of jelly from a prehistoric ocean. We know that each person on earth has a unique identification marker in their DNA. There is a set of precise and very complicated instructions for each kind of creature, intelligent information that could never possibly get there simply by the chance of nature. This knowledge is fairly recent not available to you or your associates.

11ᵗʰ There is no supernatural event in history

The historical record, the Bible, is a written word available as a reference for centuries providing the description of logical events from the beginning.

Those who insist on ignoring them even today in spite of evidence must come up with a logical explanation of our beginning. We now know the population of Earth is seven billion which relates to a beginning in thousands of years. The "Big Bang[25]" a massive explosion is the latest theory to explain a beginning and creation of all things when we choose to ignore history.

12ᵗʰ There is no Creator

We now realize that it takes a whole series of natural laws, all working together, called synergism[26], to convert a dead planet into one that supports life. This suggests the actions of some divine entity or outside force, because it demands simultaneous action.

Robert Jastrow[27] an American astronomer, physicist, cosmologist and a leading NASA scientist wrote an interesting comment. It comes from his book called "God and the Astronomers," Page 107:

"It is not a matter of another year, another decade of work, another measurement, or another theory, at this moment it seems as though science will never be able to raise the curtain

on the mystery of creation. For the scientist who has lived by his faith in the power of reason, the story ends like a bad dream. He has scaled the mountains of ignorance; he is about to conquer the highest peak; as he pulls himself over the final rock, he is greeted by a band of theologians who had been sitting there for centuries."[28]

You may wonder how I was able to obtain all this information and to question age old theories. In this the 21st century we have an amazing resource called the Internet that gives us access to nearly every book ever written. It is available in our homes, at our finger tips allowing us to easily discover the truth concerning any subject, if we are willing. What we need now are more "Free[29] Critical Thinkers"[30] willing to discover the truth.

We now know that planet earth, within the solar system, is located in the Milky Way Galaxy.[31] This galaxy is very dense with billions of stars which would normally totally obstruct our view of outer space. We are uniquely located with a "window like" opening allowing us to peer with telescopes into deep space. It is like the designer of the universe was encouraging us to explore the cosmos beyond.

However, in spite of all this information it seems that nothing has changed in 400 years. Today we continue to use these false premises for our research!

A very powerful religious system named Galileo[32] a heretic years ago for making a significant scientific discovery; this action enslaved him in his home. We find that today, a different very powerful system dictates what is taught in spite of contrary evidence. Our academic community, with government support, continues to teach theories to children at nearly every level of education. There is also extensive daily promotion of these false premises by our media proving that an old tradition is not easily changed!

This is just a short letter to bring you up to date on some of the developments over the last two millennia.

Yours Truly,

Year 2014 in the 21st Century.

References:-

1. Age of Enlightenment This page was last modified on 26 August 2014 at 19:36. 30August2014/ <http://en.wikipedia.org/wiki/Age_of_Enlightenment>

2. James Hutton This page was last modified on 13 August 2014 at 07:57 30August2014/ <http://en.wikipedia.org/ wiki/James_Hutton>

3. Charles Lyell This page was last modified on 14 August 2014 at 05:45. 30August2014/ <http://en.wikipedia.org/ wiki/Charles_Lyell>

4. Louis Pasteur This page was last modified on 27 August 2014 at 07:00. 30August2014/ <http://en.wikipedia.org/ wiki/Louis_Pasteur>

5. Biogenesis This page was last modified on 6 August 2014 at 10:53. 30August2014/ <http://en.wikipedia.org/ wiki/Biogenesis>

6. Thomas Huxley This page was last modified on 17 August 2014 at 16:41. 30August2014/ <http://en.wikipedia.org/wiki/Thomas_Henry_Huxley>

7. Abiogenesis This page was last modified on 25 August 2014 at 23:47. 30August2014/ <http://en.wikipedia.org/wiki/Abiogenesis>

8. DNA This page was last modified on 25 August 2014 at 17:43. 30August2014/ <http://en.wikipedia.org/wiki/DNA>

9. Ibid.

10. William Whewell This page was last modified on 23 August 2014 at 09:05. 30August2014/ <http://en.wikipedia.org/wiki/William_Whewell>

11. Inductive Reasoning This page was last modified on 17 August 2014 at 17:00. 30August2014/ <http://en.wikipedia.org/wiki/Inductive_reasoning>

12. Empirical Evidence This page was last modified on 26 August 2014 at 13:41. 30August2014/ <http://en.wikipedia.org/wiki/Empirical_evidence>

13. Atomic Clock This page was last modified on 9 August 2014 at 10:25. 30August2014/ <http://en.wikipedia.org/wiki/Atomic_clock>

14. Leap Second This page was last modified on 29 July 2014 at 08:27. 30August2014/ <http://en.wikipedia.org/wiki/Leap_second>

15. Hydrocarbons This page was last modified on 23 August 2014 at 20:58. 30August2014/ <http://en.wikipedia.org/wiki/Hydrocarbon>

16. Ibid.

17. Ibid.

18. Ibid

19. Age Limiting Factors Posted by R. K. Sepetjian in Fides, Scientia 27 August 2012, 30August2014/ <http://sepetjian.wordpress.com/2012/01/27/age-limiting-factors-earths-oldest-living-organism/>

20. John Woodmorappe This page was last modified on 12 July 2011, at 15:35 30August2014/ <http://creationwiki.org/John_Woodmorappe>

21. Mutations This page was last modified on 28 August 2014 at 18:50. 30August2014/ <http://en.wikipedia.org/wiki/Mutation>

22. Ibid.

23. Laws of Thermodynamics This page was last modified on 27 August 2014 at 10:57. 30August2014/ <http://en.wikipedia.org/wiki/Laws_of_thermodynamics>

24. Ibid.

25. Big Bang This page was last modified on 2 October 2014 at 22:59. 10October2014/<http://en.wikipedia.org/wiki/Big_Bang>

26. Synergism This page was last modified on 24 August 2014 at 10:03. 30August2014/ <http://en.wikipedia.org/wiki/Synergy>

27. Robert Jastrow This page was last modified on 10 June 2014 at 18:30. 30August2014/ <http://en.wikipedia.org/wiki/Robert_Jastrow>

28. Robert Jastrow, "God and the Astronomers" W.W. Norton & Company, Inc.1992, P.107

29. Free Thought This page was last modified on 21 August 2014 at 21:16 30August2014/ <http://en.wikipedia.org/wiki/Freethought>

30. Critical Thinkers This page was last modified on 26 August 2014 at 21:51. 30August2014/ <http://en.wikipedia.org/wiki/Critical_thinking>

31. Milky Way Galaxy This page was last modified on 13 August 2014 at 17:16. 30August2014/ <http://en.wikipedia.org/wiki/Milky_Way>

32. Galileo This page was last modified on 29 August 2014 at 02:48. 30August2014/ <http://en.wikipedia.org/wiki/Galileo_Galilei>

Chapter Fifteen

In Conclusion – A Love Story

A book that I read recently described how a father is concerned about his daughter. He instructed her when very young to believe in the stories contained in the Bible. It seemed that she had rejected that knowledge in favour of a new belief that included evolution taught in her school.

He planned to take her to Disney World in Florida. Once there he could talk to her trying to persuade her that what she was learning in school concerning a beginning was not the absolute truth, but instead just theories. The visit to Orlando was an interesting time for an amusement, all designed to entertain, allowing anyone to escape reality for a little while. Some of the rides referred to those evolution theories in a replica of space travel which annoyed that father. When he finally had a chance to talk to his daughter about reality including what he had taught her as a child she was not prepared to listen.

"You are spoiling our adventure in Disney World," she finally exclaimed. She then reminded father that he had sent her to school for education, telling her to learn from knowledgeable teachers. The daughter certainly had a good point.

Is it the responsibility of our children to disagree with teachers or to change an educational system? We cannot ask our children to defend what we have grown to believe as well when they are looking to educators for truth.

I think of the expression by English philosopher Edmund Burke[1] when he said,

"The only thing necessary for the triumph of evil is for good men (and women) to do nothing."

I made a recent visit to Florida as well, a result of the kindness of friends. Rather than Disney World, wife Gillian and I visited the Everglades of Florida. There I discovered a bit of reality rather than just fantasy. An airboat ride through the Everglades was the main reason for our adventure. Before that ride, I had a conversation with a man holding an alligator[2] for everyone to examine.

"You are holding a dinosaur[3]," I proceeded to explain, thinking this would shock him, or at least start some interesting conversation.

"You are correct," he exclaimed. "It's a miniature dinosaur." He then confirmed that it was a reptile[4], it continues to grow, lives about 60 years, the male reaching about 14 feet in length when fully grown. To my surprise, even children I talked to later, living in these tropical conditions knew that alligators were small dinosaurs with a different name.

However, dinosaurs[5] reach a much larger size even up to 150 feet long with tail.

Why are dinosaurs so much larger than today's reptiles including Crocodiles, Geckoes, Lizards and Turtles? Does science ever answer that question?

156

Do finding dinosaur bones in every direction suggest a tropical climate everywhere at some time on our planet because this is where they live?

Does the discovery of dinosaur bones suggest recent burial, even if petrified, rather than "millions of years" old without total decay?

It seems that a simple application of Deductive Reasoning[6] could answer those questions and more! Remembering my grade school education, I decided to apply some of the logic learned there. The ABC's of Deductive Reasoning[7], if A equals B, and B equals C, therefore A equals C!

A dinosaur is a reptile, an alligator is a reptile, and therefore an alligator is a dinosaur. The reptile[8] may have a different name now but it is still a reptile. If you examine the skeleton of an alligator, it resembles a dinosaur[9], only much smaller! This may lead to research why dinosaurs are not extinct just smaller rather than finding how they evolved into birds because reptiles also lay eggs.

I can obtain other truths also by application of this logic. The alligator reaches a length of 14 feet in 60 years; a dinosaur reaches a length of about 150 feet. Alligators and dinosaurs are both reptiles that never stop growing. Therefore, a dinosaur must be ten times older, about 600 years old!

Why do reptile's now living stop growing at about 14 feet or do not live for 600 years?

Historical records tell us that humans once lived up to 900 years. Is it logical Deductive Reasoning[10] that dinosaurs and humans lived much longer during a much different climate and set of conditions protected by a canopy above as described in the Bible. That book also provides an excellent description of a giant creature in Job 40:15-19 called a "behemoth!" A previous chapter gives a detail description of this giant creature.

May I suggest that the father mentioned previously decide to take his daughter to the Everglades of Florida rather than Disney World if he needs evidence to support his belief!

The chapters of this book conclude with the word **"therefore"** the final stage of Deductive Reasoning suggesting that this is an ideal method to use in discovering the truth, rather than Inductive Reasoning[11] that leads to more theoretical ideas.

The father had a very good reason to be concerned with what his daughter is learning because he loves her. Love is not something you can put in a test tube for scientific analysis or find in a theory. Rather, the "Evolution Theories" teach "survival of the fittest[12]," no consideration for others, only selfish direction, and a "me only" attitude.

Will this teaching lead to a peaceful future for our children or do we have contrary evidence from past history?

History witnessed that selfish attitude producing negative results during the last world war. A whole nation of children taught they were a master race[13] at the highest level of the evolutionary chain superior to any other, with a right to impose their beliefs on others. Is this happening again in our schools? Battles we won then, but did we win the war?

The "survival of the fittest[14]" idea suggests a constant struggle for humanity while yet another group of people consider themselves the "fittest." I prefer the ideas expressed by the Judeo-Christian message that love for humanity is a preferred method. This message includes cooperation with a Creator who taught us about peaceful harmony for humanity, a world where love, forgiveness, and respect is freely given.

I learned years ago that when a business starts to fail it is time to get back to the "basics" of what made that business successful. We hear of violent crimes daily and an increasing breakup of the family home. Isn't it time we got back to the basics?

There is another well known belief system that teaches those "basics" for peace and harmony.

The ten commandments of the Judeo-Christian Bible include loving respect for a Creator God with honour and respect for others.[15] The story continues in the New Testament with that theme where it states, "God is love."[16]

When someone asked Jesus what the two greatest commandments were;

"Jesus replied: 'Love the Lord your God with all your heart and with all your soul and with all your mind. This is the first and greatest commandment. And the second is like it: 'Love your neighbour as yourself.' All the Law and the Prophets hang on these two Commandments[17]."

That we should even love our enemies[18] is an additional teaching, much different than the evolutionary theories with no moral principles or respect for others!

What is the future for our children when the "basis" for love may not be part of the agenda in our schools or elsewhere?

159

Is it time that we got back to reality, to basic beliefs instead of only teaching theories. Those Judeo-Christian beliefs were the ones that built a free democratic society, a respect for others that many fought and died to protect. There is much evidence that a spiritual belief transforms us, the following is one example:

John Bunyan[19] was a foul-mouthed "tinker" (one who repairs old pots and pans) with very little formal education living in the 17th century. A spiritual awakening sometimes called being "born again[20]" changed him. He is the author of Pilgrims Progress[21] one of the most famous allegories ever written to explain his spiritual journey.

For each one of us it is a personal journey, a free choice, no other person or organization can do that for each one of us.

In the book of Genesis, a Creative Spirit transformed a dead planet into a living one. There is ample evidence that the same Spirit can transform our dead Spirit into a living one, *with a future[22]*, as it did for John Bunyan[23]. This is a free choice for each of us but it certainly helps when we learn about it somewhere. A person may never hear about it if not taught or mentioned in the home or some other centre for learning.

The concept of a Spirit taking on a "bodily form" like us may be difficult to accept or understand as it is so foreign to our experience and way of thinking. The close followers of Jesus also had trouble with this concept, especially Thomas, sometimes called doubting Thomas. He asked Jesus a direct question:

"Thomas said to him, Lord, we don't know where you are going, so how can we know the way[24]? Jesus answered, I am the way and the truth and the life. No one comes to the Father except through me[25]"

Quite a statement for us to reflect on!

The historical record called the Bible reveals that when the Creative Spirit became "human bodily flesh" He also gave us the responsibility to help our neighbour around the world in that search for the truth. The Great Commission[26] is the name of that task. The following poem by Henry Wadsworth Longfellow[27] includes this call to action:

PSALM OF LIFE

WHAT THE HEART OF THE YOUNG MAN
SAID TO THE PSALMIST
TELL me not, in mournful numbers,
Life is but an empty dream! —
For the soul is dead that slumbers,
And things are not what they seem.
Life is real! Life is earnest!
And the grave is not its goal;
Dust thou art, to dust returnest,
Was not spoken of the soul.
Not enjoyment, and not sorrow,
Is our destined end or way;
But to act, that each to-morrow
Find us farther than to-day.
Art is long, and Time is fleeting,
And our hearts, though stout and brave,
Still, like muffled drums, are beating
Funeral marches to the grave.
In the world's broad field of battle,
In the bivouac of Life,

Be not like dumb, driven cattle!
Be a hero in the strife!
Trust no Future, howe'er pleasant!
Let the dead Past bury its dead!
Act,— act in the living Present!
Heart within, and God o'er head!
Lives of great men all remind us
We can make our lives sublime,
And, departing, leave behind us
Footprints on the sands of time;
Footprints, that perhaps another,
Sailing o'er life's solemn main,
A forlorn and shipwrecked brother,
Seeing, shall take heart again.
Let us, then, be up and doing,
With a heart for any fate;
Still achieving, still pursuing,
Learn to labor and to wait.

A final therefore! Scientific knowledge provides evidence for the accuracy of the historical record of a beginning, provided for us in a book called Genesis, describing those events!

To understand our world we must include knowledge of both history and science, giving us a very different perspective than if we decide to exclude one of them. When we make the effort to research both, with an open mind, we find much more information and evidence to arrive at logical conclusions.

In addition, as scientific knowledge progresses, we discover a display of even more amazing creative designs. Can this possibly happen without an intelligent designer?

The alternative to this thinking is Evolution Theories. This belief system includes several theoretical ideas for the creation of all matter, seen and unseen. This idea only considers the selections of nature which has no prior knowledge of the final product, not using any planned design or intelligent designer, all happening simply by chance, over eons of time.

This is the reason for me stating, the choices we make determine our destiny; choose well my friends!

References:-

(All biblical references are from the New International Version)

1. Edmund Burke This page was last modified on 27 July 2014 at 18:39. 19August2014/ <http://en.wikipedia.org/wiki/Edmund_Burke>

2. Alligator This page was last modified on 21 June 2014 at 10:14. 19August2014/ <http://en.wikipedia.org/wiki/Alligator>

3. Dinosaur This page was last modified on 10 August 2014 at 18:03. 19August2014/ <http://en.wikipedia.org/wiki/Dinosaur>

4. Reptile This page was last modified on 18 August 2014 at 14:15. 19August2014/ <http://en.wikipedia.org/wiki/Reptile>

5. Ibid.

6. Deductive Reasoning This page was last modified on 4 August 2014 at 15:29. 19August2014/ <http://en.wikipedia.org/wiki/Deductive_reasoning>

7. Ibid.

8. Ibid.

9. Ibid.

10. Ibid.

11. Inductive Reasoning This page was last modified on 14 September 2014 at 14:01. 23September2014/ <http://en.wikipedia.org/wiki/Inductive_reasoning>

12. Survival of the Fittest This page was last modified on 13 August 2014 at 08:45. 19August2014/ <http://en.wikipedia.org/wiki/Survival_of_the_fittest>

13. Master race This page was last modified on 9 September 2014 at 14:21. 12September2014/ <http://en.wikipedia.org/ wiki/Master_race>

14. Ibid.

15. Deuteronomy 5 (Ten Commandments)

16. 1 John 4:8 to 16

17. Matthew 22:37 to 40

18. Luke 6: 27 & 28

19. John Bunyan This page was last modified on 9 August 2014 at 08:24. 19August2014/ <http://en.wikipedia.org/wiki/John_Bunyan>

20. Born Again This page was last modified on 9 September 2014 at 20:39. 27September2014/ <http://en.wikipedia.org/wiki/Born_again_(Christianity)>

21. Pilgrims Progress This page was last modified on 12 August 2014 at 21:59. 19August2014/ <http://en.wikipedia.org/wiki/The_Pilgrim%27s_Progress>

22. John 14: 2 "In my Father's house are many rooms; if it were not so, I would have told you. I am going there to prepare a place for you."

23. Ibid.

24. John 14:5

25. John 14:6

26. Matthew 28: 16 to 20

27. Henry Wadsworth Longfellow This page was last modified on 24 August 2014 at 18:30. 19August2014/ <http://en.wikipedia.org/wiki/Henry_Wadsworth_Longfellow>

Contact the Author-

Stanley Bachelder

PO Box 173 Caledon East

Caledon East ON

L7C 3L9

www.stanbachelder.com

-or-

info@stanbachelder.com

-or-

info@allowmetothink.com

"A lie doesn't become truth, wrong doesn't become right & evil doesn't become good, just because it's accepted by a majority."

Author Unknown

Bibliography...

A

Abiogenesis
<http://en.wikipedia.org/wiki/Abiogenesis>
Adler, John with John Carey: (Archaeology) *Is Man a Subtle Accident,*
Newsweek, Vol.96, No.18 (November 3, 1980, p.95)
Agassiz, Prof. J (Natural History) "Professor Agassiz on the Origin of Species,"
American Journal of Science 30 (June 1860):143-47, 149-50
Age of Enlightenment
<http://en.wikipedia.org/wiki/Age_of_Enlightenment>
Alligator
<http://en.wikipedia.org/wiki/Alligator>
Antediluvian
<http://en.wikipedia.org/wiki/Antediluvian>
Apollo 11
<http://en.wikipedia.org/wiki/Apollo_11>
Article obtained Nov. 10, 2013 from United States Environmental Agency.
Article called Schweitzer's Dangerous Discovery from Discover Magazine,
April 2006 Issue
Atomic Clock
<http://en.wikipedia.org/wiki/Atomic_clock>

B

Bachiler, Stephen
<http://en.wikipedia.org/wiki/Stephen_Bachiler>
Bar Mitzvah
<http://en.wikipedia.org/wiki/Bar_and_Bat_Mitzvah>
Behe, Michael
<http://en.wikipedia.org/wiki/Michael_Behe>

Consistency
<http://en.wikipedia.org/wiki/Consistency>
Cosmic Dust
<http://en.wikipedia.org/wiki/Cosmic_dust>
Cracking Tower
<http://en.wikipedia.org/wiki/Catalytic_cracking>
Cradle of Civilization
<http://en.wikipedia.org/wiki/Cradle_of_civilization>
Cranmer, Thomas
<http://en.wikipedia.org/wiki/Thomas_Cranmer>
Critical Thinkers
<http://en.wikipedia.org/wiki/Critical_thinking>

D

Darwin's Black Box
<http://en.wikipedia.org/wiki/Darwin%27s_Black_Box>
Darwin, Charles
<http://en.wikipedia.org/wiki/Charles_Darwin>
Darwinism
<http://en.wikipedia.org/wiki/Darwinism>
Dawkins, Richard
<http://en.wikipedia.org/wiki/Richard_Dawkins>
Dead Sea Scrolls
<http://en.wikipedia.org/wiki/Dead_Sea_scrolls>
Deductive Reasoning
<http://en.wikipedia.org/wiki/Deductive_reasoning>
Deep Time
<http://en.wikipedia.org/wiki/Deep_time>
Deluge
<http://en.wikipedia.org/wiki/Deluge>
Deuteronomy 5 and 6
Diamond
<http://en.wikipedia.org/wiki/Diamond>
Dinosaur
<http://en.wikipedia.org/wiki/Dinosaur>
DNA
<http://en.wikipedia.org/wiki/DNA>
Dragons
<http://en.wikipedia.org/wiki/Dragon>

Drumlins
<http://en.wikipedia.org/wiki/Drumlin>

E
Earth's Magnetic Field
<http://en.wikipedia.org/wiki/Earth%27s_magnetic_field>

Empirical Evidence
<http://en.wikipedia.org/wiki/Empirical_evidence>
Entropy
<http://en.wikipedia.org/wiki/Entropy>
Eocene Epoch
<http://www.encyclopedia.com/topic/Eocene_epoch.aspx>
Eugenics
<http://en.wikipedia.org/wiki/Eugenics>
Expelled-No Intelligence Required
<http://en.wikipedia.org/wiki/Expelled:_No_Intelligence_Allowed>

F
Fertile Crescent
<http://en.wikipedia.org/wiki/Fertile_Crescent>
Firmament
<http://en.wikipedia.org/wiki/Firmament>
First Law of Thermodynamics,
<http://en.wikipedia.org/wiki/First_law_of_thermodynamics>
Flagella Bacteria (Flagellum)
<http://en.wikipedia.org/wiki/Flagellum>
Fossil Fuels
<http://en.wikipedia.org/wiki/Fossil_fuel>
Freethought
<http://en.wikipedia.org/wiki/Freethought>
From Darwin to Hitler
<http://en.wikipedia.org/wiki/From_Darwin_to_Hitler>

G
Galilei, Galileo,
<http://en.wikipedia.org/wiki/Galileo_galilei>
Geological Time Scale
<http://en.wikipedia.org/wiki/Geologic_time_scale>

Glaciations

<http://en.wikipedia.org/wiki/Timeline_of_glaciation>

Glacial Periods

<http://en.wikipedia.org/wiki/Glacial_period>

Goldilocks Principle

<http://en.wikipedia.org/wiki/Goldilocks_Principle>

Gould, Stephen Jay (Geology) *(Professor of Geology and Palaeontology, Harvard University), "Is a new and general theory of evolution emerging?" Palaeontology, vol. 6(1), January 1980, p. 127*

Great Lakes Basin

<http://en.wikipedia.org/wiki/Great_Lakes_Basin>

Goldschmidt, Prof. (Chemistry) *R PhD, DSc Prof. Zoology, University of Calif. In Material Basis of Evolution Yale Univ. Press*

Green, David and Robert Goldberger (Biology) (1967) *Molecular Insights into the Living Process New York: Academic Press, pp. 406-407. (David Green is a biochemist at the University of Wisconsin. Robert Goldberger is a biochemist at the National Institute of Health)*

H

Hebrew Poetry

<http://en.wikipedia.org/wiki/Biblical_poetry>

Holy Roman Empire

<http://en.wikipedia.org/wiki/Holy_Roman_Empire>

Hoyle, Sir Fred (Astronomy-Cosmology) *(English astronomer, Professor of Astronomy at Cambridge University), as quoted in "Hoyle on Evolution." Nature, vol. 294, 12 Nov. 1981, p. 105*

Hutton, James

<http://en.wikipedia.org/wiki/James_Hutton>

Huxley, Thomas

<http://en.wikipedia.org/wiki/Thomas_Henry_Huxley>

Hydrocarbons

<http://en.wikipedia.org/wiki/Hydrocarbon>

I

Ice Age

<http://en.wikipedia.org/wiki/Ice_age>

Index Fossils

<http://en.wikipedia.org/wiki/Index_fossil>

Inductive Reasoning

<http://en.wikipedia.org/wiki/Inductive_reasoning>

Luke 6: 27 & 28
Lyell, Charles
<http://en.wikipedia.org/wiki/Charles_Lyell>

M

Mark 13:26
Mary I of England
<http://en.wikipedia.org/wiki/Mary_I_of_England>
Masoretic Text
<http://en.wikipedia.org/wiki/Masoretic_Text>
Master race
<http://en.wikipedia.org/wiki/Master_race>
Matthew 22:37 to 40
Matthew 28: 16 to 20
Mesopotamia
<http://en.wikipedia.org/wiki/Mesopotamia>
Megalosaurus
<http://en.wikipedia.org/wiki/Megalosaurus>
Memorial Diamonds
<http://www.memorial-diamonds.com
Methane Gas
<http://en.wikipedia.org/wiki/Methane>
Milky Way Galaxy
<http://en.wikipedia.org/wiki/Milky_Way>
Moon
<http://en.wikipedia.org/wiki/Moon>
Moses (Royal Courts of Egypt)
<http://en.wikipedia.org/wiki/Moses>
Mount Ararat
<http://en.wikipedia.org/wiki/Mount_Ararat>
Mutation Rate
<http://en.wikipedia.org/wiki/Mutation_rate>
Mutations
<http://en.wikipedia.org/wiki/Mutation>

N

Natural Selection
<http://en.wikipedia.org/wiki/Natural_selection>
Niagara Gorge
<http://en.wikipedia.org/wiki/Niagara_Gorge>

Noah's Ark Found in Turkey? Published April 28, 2010,
<http://news.nationalgeographic.com/news/2010/04/100428-noahs-ark-found-in-turkey-science-religion-culture/>
Nonconformist
<http://en.wikipedia.org/wiki/Nonconformist>

O
Oral Tradition
<http://en.wikipedia.org/wiki/Oral_tradition>
Ordovician Age
<http://en.wikipedia.org/wiki/Ordovician>
O'Rourke, J. *in.the American Journal of Science, January 1976, Volume 276, Page 51*
On the Origin of Species by Means of Natural Selection, or the Preservation of Favoured Races in the Struggle of Life, 1859, Chapter VI, Difficulties, Page 85
Owen, Sir Richard
<http://en.wikipedia.org/wiki/Richard_Owen>

P
Pasteur, Louis
<http://en.wikipedia.org/wiki/Louis_Pasteur>
Pilgrims Fathers
<http://en.wikipedia.org/wiki/Pilgrims_(Plymouth_Colony)>
Pilgrims Progress
<http://en.wikipedia.org/wiki/The_Pilgrim%27s_Progress>
Playfair, John
<http://en.wikipedia.org/wiki/John_Playfair>
Principles of Geology
<http://en.wikipedia.org/wiki/Principles_of_Geology>
Prophecies, 60
<http://www.bible.ca/b-prophecy-60.htm>
Proxima Centauri
<http://en.wikipedia.org/wiki/Proxima_Centauri>
Puritan
<http://en.wikipedia.org/wiki/Puritan>

R
Reich
<http://en.wikipedia.org/wiki/Reich>

Reptile

<http://en.wikipedia.org/wiki/Reptile>

River Deltas

<http://en.wikipedia.org/wiki/River_delta>

Rock Cycle

<http://en.wikipedia.org/wiki/Rock_cycle>

S

Santayana, George

<http://en.wikipedia.org/wiki/George_Santayana>

Schweitzer, Mary Highby

<http://en.wikipedia.org/wiki/Mary_Higby_Schweitzer>

Scientific Method

<http://en.wikipedia.org/wiki/Scientific_method>

Scientifically Unproven

<http://www.wicwiki.org.uk/mediawiki/index.php/Scientifically_Unproven>

Scribes

<http://en.wikipedia.org/wiki/Scribe>

Search for Extraterrestrial Intelligence-SETI

<http://en.wikipedia.org/wiki/SETI>

Second Law of Thermodynamics

<http://en.wikipedia.org/wiki/Second_law_of_thermodynamics>

Second Speech on Foote's Resolution, U.S. Senate, 26 Jan. 1830,
12August2014/ <http://izquotes.com/quote/311624>

Secular Humanism

<http://en.wikipedia.org/wiki/Secular_humanism>

Separation of Church & State

<http://en.wikipedia.org/wiki/Separation_of_church_and_state>

Smith, Wolfgang (Physics) in his book "Teilhardism and the New Religion: A
Thorough Analysis of the Teachings of Pierre Teilhard de Chardin", Tan Books
& Pub. Inc: Rockford (USA), 1988 p: 1

Social Darwinism

<http://en.wikipedia.org/wiki/Social_Darwinism>

Spanish Armada

<http://en.wikipedia.org/wiki/Spanish_Armada>

Speed of Light

<http://en.wikipedia.org/wiki/Speed_of_light>

Spencer, Herbert

<http://en.wikipedia.org/wiki/Herbert_Spencer>

Spinoza, Baruch
< http://en.wikipedia.org/wiki/Baruch_Spinoza>
Sumeria
<http://en.wikipedia.org/wiki/Sumeria>
Survival of the Fittest
<http://en.wikipedia.org/wiki/Survival_of_the_fittest>
Symbiosis
<http://en.wikipedia.org/wiki/Symbiosis>
Synergistic
<http://en.wikipedia.org/wiki/Synergy>

T

Tahmisian, Dr. T. N. (Atomic Energy Commission, USA) in "The Fresno Bee,"
August 20, 1959. As quoted by N. J. Mitchell, Evolution and the Emperor's
New Clothes, Roydon Publications, UK, 1983, title page.
"The Greatest Show on Earth" Chapter 4 called "Silence and Slow Time" page 85
The Privileged Planet a DVD produced by Illustra Media by Jay Richards and
Guillermo
Gonzales, Regnery Publishing
Theistic Evolution
<http://en.wikipedia.org/wiki/Theistic_evolution>
Tidal Friction
<http://en.wikipedia.org/wiki/Tidal_friction>
Tides
<http://en.wikipedia.org/wiki/Tide>
Time Scale
<http://en.wikipedia.org/wiki/Time_scale>
Torah
<http://en.wikipedia.org/wiki/Torah>
Transitional Fossils -Missing Links
<http://en.wikipedia.org/wiki/Transitional_fossil>
Type of Rock (List of Rock Types)
<http://en.wikipedia.org/wiki/List_of_rock_types>

U

Uniformitarianism
<http://en.wikipedia.org/wiki/Uniformitarianism>

V

Vernix Caseosa

<http://en.wikipedia.org/wiki/Vernix_caseosa>

Very Good, Genesis 1:31 New International Version "God saw all that he had made, and it was very good. And there was evening, and there was morning--the sixth day."

W

Webster, Daniel

<http://en.wikipedia.org/wiki/Daniel_Webster>

Webster, Daniel © 2014 Goodreads Inc.,

<http://www.goodreads.com/author/quotes/31180.Daniel_Webster>

Webster, Daniel Quotes

<http://quotes.liberty-tree.ca/quote_blog/Daniel.Webster.Quote.73C7>

Whewell, William

<http://en.wikipedia.org/wiki/William_Whewell>

Woodmorappe, John

<http://creationwiki.org/John_Woodmorappe>

World Population

<http://en.wikipedia.org/wiki/World_population>

Y

Y-Chromosome

<http://en.wikipedia.org/wiki/Y_chromosome>